CLIMATE CHANGE
The Facts

CLIMATE CHANGE
The Facts

Edited by Alan Moran

With contributions from
John Abbot & Jennifer Marohasy
Andrew Bolt ~ Robert M Carter
Rupert Darwall ~ James Delingpole
Christopher Essex ~ Stewart W Franks
Kesten C Green & J Scott Armstrong
Donna Laframboise ~ Nigel Lawson
Bernie Lewin ~ Richard S Lindzen
Ross McKitrick ~ Alan Moran
Patrick J Michaels ~ Joanne Nova
Garth W Paltridge ~ Ian Plimer
Willie Soon ~ Mark Steyn
and Anthony Watts

STOCKADE
BOOKS

First published by the Institute of Public Affairs, Melbourne, Victoria

Published in 2015 by
Stockade Books
Box 30, Woodsville, New Hampshire 03785

Printed and bound in the Province of Quebec, Canada

ISBN: 978-0-9863983-0-8

Contents

Contributors

John Abbot
Senior Research Fellow at Central Queensland University.

J. Scott Armstrong
University of Pennsylvania, Ehrenberg-Bass Institute, Australia.

Andrew Bolt
Columnist with the *Herald Sun, Daily Telegraph*, and the *Advertiser*, and host of Channel Ten's *The Bolt Report*.

Robert M. Carter
Emeritus Fellow and Science Policy Advisor at the Institute of Public Affairs, science adviser at the Science and Public Policy Institute; chief science adviser for the International Climate Science Coalition (ICSC), and former Professor and Head of the School of Earth Sciences at James Cook University.

Rupert Darwall
Rupert Darwall is the author of *The Age of Global Warming, A History* published in 2014 and has also written for the *Wall Street Journal* and the *Financial Times*. He holds a degree in Economics and History from Cambridge University, and in 1993 was a Special Adviser to Her Majesty's Treasury.

James Delingpole

Executive editor for the London branch of Breitbart.com; author of *Watermelons: How Environmentalists are Killing the Planet, Destroying the Economy and Stealing Your Children's Future* (published in Australia as *Killing the Earth to Save It*).

Christopher Essex

Chairman of the Permanent Monitoring Panel on Climate, World Federation of Scientists and Professor and Associate Chair at the Department of Applied Mathematics, University of Western Ontario.

Stewart W. Franks

Foundation Chair of Environmental Engineering, University of Tasmania.

Kesten C. Green

Senior Research Associate of the Ehrenberg-Bass Institute, University of South Australia, and Senior Lecturer at University of South Australia

Donna Laframboise

Journalist, author of *The Delinquent Teenager Who Was Mistaken for the World's Top Climate Expert*; former member of the Board of Directors of the Canadian Civil Liberties Association 1993-2001.

Nigel Lawson

Former British Conservative politician and journalist; member of the House of Lords; author of *An Appeal to Reason: A Cool Look at Global Warming*.

Bernie Lewin

Author of climate change blog, *Enthusiasm, Scepticism and Science* (enthusiasmscepticismscience.wordpress.com)

CONTRIBUTORS

Richard S. Lindzen

Atmospheric physicist, former professor of Meteorology at Massachusetts Institute of Technology 1983-2013; Distinguished Senior Fellow, Center for the Study of Science at the Cato Institute; lead author of Chapter 7, 'Physical Climate Processes and Feedbacks,' of the IPCC's *Third Assessment Report* (2001).

Jennifer Marohasy

Adjunct Research Fellow in the Centre for Plant and Water Science at Central Queensland University; foundation member of the Australian Environment Foundation, former Senior Fellow at the Institute of Public Affairs 2004-2009; author of *Myth and the Murray: Measuring The Real State of the River Environment.*

Ross McKitrick

Professor of Economics at University of Guelph, co-author of *Taken By Storm: The Troubled Science, Policy and Politics of Global Warming* and author of *Economic Analysis of Environmental Policy.*

Patrick J. Michaels

Director of the Center for the Study of Science at the Cato Institute, Washington DC; research professor of Environmental Sciences at University of Virginia 1980-2007; author of *Shattered Consensus: The True State of Global Warming* and *Climate Coup: Global Warming's Invasion of Our Government and Our Lives.*

Alan Moran

Manager of Regulation Economics, former Director of the Deregulation Unit at the Institute of Public Affairs. Director of the Australian Office of Regulation Review and Deputy Secretary Energy in the Victorian Department of Minerals and Energy.

Jo Nova
Science writer with columns published in *The Spectator* and *The Australian*, author of *The Skeptic's Handbook* (2009).

Garth W. Paltridge
Emeritus Professor and Honorary Research Fellow at the Institute of Antarctic and Southern Oceans Studies (IASOS), University of Tasmania; visiting fellow at the Australian National University and fellow of the Australian Academy of Science. Former chief research scientist with the CSIRO Division of Atmospheric Research and former CEO of the Antarctic Cooperative Research Centre. Author of *The Climate Caper: facts and fallacies of global warming* (2009).

Ian Plimer
Emeritus Professor of Earth Sciences at the University of Melbourne, Professor of Mining Geology at the University of Adelaide, and author of *Not For Greens* (2014).

Willie Soon
Independent scientist.

Mark Steyn
Political commentator and author of many books including *After America: Get Ready for Armageddon* (2011) and *America Alone: The End of the World as We Know It* (2006).

Anthony Watts
Retired American Meteorology Society certified television meteorologist, author of 'Is the US Surface Temperature Record Reliable?' published the Heartland Institute (2009).

Introduction

Alan Moran

Prompted by successive reports of the Intergovernmental Panel on Climate Change (IPCC), the issue of human induced climate change has become a dominant theme of world politics. This is especially so in Australia where it was famously called the greatest moral challenge of our time by Kevin Rudd. The issue was pivotal to Mr Rudd's replacement in 2010 as prime minister by Julia Gillard, his subsequent restoration to that position and his loss to Tony Abbott in the election of 2013.

The book is divided into three parts. Part one examines the science of climate change.

Ian Plimer examines both the science and politics behind climate change. He notes that many Western governments have a politically popular ideology involving human emission increases in carbon dioxide (CO_2) bringing warming, possible catastrophic 'tipping points' and a need to phase out fossil fuels as the only means of stopping this. He dismisses the possibility of devastating consequences, drawing from geological history, and points to the adverse economic outcomes of attempts to drastically reduce fossil fuel based energy usage.

Patrick Michaels examines the contrast between the predictions of the IPCC and outcomes. He details and demolishes the manifold excuses for this put forward by Obama adviser, formerly Club of Rome alarmist,

John Holdren, and other IPCC faithful. He argues that a growing body of academic literature now suggests the IPCC models overstate the likely future increases in global temperatures.

Richard Lindzen demonstrates that the climate is relatively insensitive to increases in greenhouse gases, and that in any event a warmer world would have a similar variability in weather to that we have always seen. He deplores the acquiescence of scientists to the exaggerated concerns over climate change.

Willie Soon explodes the myth that 97 per cent of scientists regard human induced global warming as both likely and serious. He shows that it is the sun, which like all stars has different levels of radiative intensity, that is overwhelmingly the cause of Earth's climatic variations.

Robert Carter shows that any human effect on climate is trivial compared to natural variation, and that there is no evidence the next 50 years will bring human induced warming. He maintains that the appropriate role for governments is to prepare for natural changes in climate that might occur.

John Abbot and Jennifer Marohasy argue that a century ago, Australian meteorologists focused on solar, lunar and planetary cycles to understand climate but, like other such institutions around the world, Australia's Bureau of Meteorology now largely seeks to simulate climate largely independently of extraterrestrial influences. Marohasy and Abbot use rainfall as an example of how outcomes have borne little relationship to the forecasts. They regard the shift in emphasis to have stemmed from a change in science funding towards reliance on governments with the political baggage this brings.

Part two develops these themes and explores the politics and economics of climate change.

Nigel Lawson notes that UK climate change secretary Ed Davey and Prince Charles were among those who vilify their opponents with the 'denier' label (and recently the UK prime minister sacked climate change sceptic Owen Paterson as secretary for the environment). Lawson explores the dire economic implications of trying to cease the use of fossil fuels. He also demonstrates the trivial effects of the warming that is predicted and discounts its claimed negative effects, noting that scientific developments mean we are far less hostage to climate shifts than in previous eras.

My own chapter sets the context of the economic debate by examining the costs of taking action (which are considerable and massively understated by the IPCC) and any benefits of doing so (which are slender and overstated by the IPCC). And the chapter notes that any gains rely on the unlikely event of a comprehensive international agreement.

James Delingpole notes how the climate believers so often accuse sceptics of lack of credentials. He delves into the qualifications of the major promoters of the climate scare in the UK and Australia and finds wall-to-wall English literature graduates. When confronted by genuine scientists who dissent from their own view, they invariably suggest the dissenting opinions are dictated by bribes from 'Big Oil'. And yet it is so often vested interests, like Munich Re, that promote the notion of dangerous climate change. The BBC's denial of platforms to sceptical scientists and the hounding of the eminent Professor Bengtsson from Lord Lawson's Global Warming Foundation illustrates how the establishment seeks to close down debate.

Garth Paltridge recaps the issues confronting meteorologists in 1970 when they first contemplated climate forecasting. These were clouds, solar balance, and oceanic behaviour. He notes that our knowledge has hardly advanced but that the IPCC tables inaccurate reports which receive little questioning from scientists even though scepticism is supposedly central to the profession. And, as *Climategate* showed, some scientists have crossed the boundary into 'post modern science'. He sees considerable backlash on the credibility of all scientists should global warming fail to eventuate.

Jo Nova points out that, globally, renewables investment has reached $359 billion annually while the EU says it will allocate 20 per cent of its budget to climate related spending. All this is based on a naive modelling of the atmosphere that employs amplifications of water vapour's influence by enhanced levels of carbon dioxide. She estimates money dedicated to promoting the global warming scare is maybe one hundred fold the funding to sceptics. She shows how the purveyors of human induced global warming use their funding to denigrate opponents and to hide contrary evidence.

Kesten Green and J. Scott Armstrong test the predictive validity of the global warming hypothesis and find it wanting. They point out that many other alarms have been raised over the past 200 years, none of which have proved to have substance. Most of the alarms that led governments into taking action actually created harm and none provided benefits.

Part three explores the climate change movement and the development of the international institutional framework and the growing disconnect from science and scientific observation that characterises the public debate.

Rupert Darwall reviews the farce of the 2009 Copenhagen conference and the subsequent mini-conferences. He notes the veto imposed on costly actions by the increasingly important third world nations, contrasting this with the revolutionary outcome that the IPCC operatives are planning to emerge from Paris in 2015.

Ross McKitrick addresses the trials he and Steve McIntyre went through in puncturing the late twentieth century myth that temperatures are now higher than at any time in the past millennium. Having been pilloried for bucking the establishment and undermining the IPCC's poster-child 'hockey stick' graph, the accuracy of their analysis has finally prevailed.

Donna Laframboise notes the scandalous attribution of Nobel Prize status to all involved in the IPCC. She traces qualifications of senior and lead authors and finds them often to be activists with no significant credentials.

Mark Steyn's essay 'Ship of Fools' demonstrates how environmental activist, Professor Chris Turney inadvertently parodied Douglas Mawson's Antarctic expedition. Turney had expected to see a path to the Pole cleared for his ship by global warming. Instead, Turney's *Guardian* backed expedition had to be rescued from expanding ice. A genuine scientist, as Turney claims to be, should have realised that Antarctic ice is expanding not increasing.

Christopher Essex shows that we cannot have intelligent public discourse until people set aside appeals to expertise and develop some expertise of their own. In the absence of that, we are reduced to debating the virtue of persons and counting heads instead of considering nature. Meanwhile we yield power to smooth talkers who use the word 'science' as a prop to frighten us into pursuing their agendas.

Bernie Lewin traces the antecedentaries of the current IPCC and how scientists, many of them genuinely seeking to uncover man's impact on climate, were hijacked by developing country interests and activists into becoming frontmen for a politicised UN agency.

Drawing heavily upon Karl Popper's theories that scientific material should be subject to constant examination and should be falsifiable, Stewart Franks points to the many phenomena of climate change that the increase in greenhouse gases both failed to predict and fail to explain.

Anthony Watts illustrates the trivial increase in global temperature that has occurred over the past century (with no increase in the past eighteen years). He notes the change in language by alarmists from 'warming' to 'climate change' in an attempt to substitute extreme climate events for the now non-existent warming trend. His examination of these extreme events—snow, storms, rainfall—shows an absence of evidence to indicate marked changes over recent decades.

Andrew Bolt disinters the graveyards of failed forecasts by climate doomers. These include the spectacular forecasts by Tim Flannery that Australian cities would run out of fresh water, by Professor Hoegh-Guldberg that the Great Barrier Reef would die, by Professor Karoly that the Murray Darling would see increasing drought, by the UK Met Office that warming would resume, and by Ross Garnaut and Al Gore that hurricanes would increase. He considers the warmistas monumental failures are finally denting the faith in them by the commentariat and politicians.

The
science of
climate change

1   The science and politics of climate change

Ian Plimer

We derive scientific evidence from measurement, observation, and experiment. Evidence must be repeatable and collected over and over again. Computers do not generate evidence: they analyse evidence that should have been repeated and validated. On the basis of the evidence and analysis of evidence, an explanation is given. This explanation is a scientific theory and must be in accord with other validated evidence from diverse sources (this is known as the coherence criterion in science). Unlike in law, there is no inadmissible evidence in science. Science is underpinned by practitioners who must be sceptical of the methodology used to collect evidence, the analysis of evidence, and the conclusions based on the evidence. On the basis of new evidence, scientists must always be prepared to change their opinions.

Science bows to no authority, is not based on a consensus, and is in a constant state of flux. No great advance in science has been made by consensus: advances have been made by individuals paddling upstream. If a scientific theory is not in accord with validated evidence, then the theory must be abandoned and reconstructed. It is scepticism that underpins science, not the comfort of consensus.

The theory of human-induced global warming is not science because research is based on a pre-ordained conclusion, huge bodies of evidence are ignored, and the analytical procedures are treated as evidence. Furthermore, climate 'science' is sustained by government research grants. Funds are not available to investigate theories that are not in accord with government ideology.

Many Western governments have a politically popular ideology that argues that:

i. There is an increase in emissions of carbon dioxide (CO_2) by human activities;

ii. The increased CO_2, a greenhouse gas, will lead to ever increasing global warming;

iii. There will be tipping points, sea level rises, extinctions and ocean acidification;

iv. Climate change will be irreversible and that human emissions of CO_2 must be reduced or stopped as soon as possible; and

v. In order to stop climate change, energy sources need to be shifted from coal, gas and oil to wind, solar, tidal and biomass.

i. There is an increase in emissions of carbon dioxide by human activities

Point (i) is correct. These emissions derive mainly from the developing world and the understandable desire of its people to reach the same standard of living as the Western middle class. It was coal in the Industrial Revolution that originally led to the rise of the middle class in the West. Now the new industrial revolution in China, India and East Asia is causing the largest migration of humans that has ever occurred, the rise of the middle class in these nations, and the use of steel and electricity, both of which derive from coal. The very slight increase in atmospheric CO_2 has led to a slight greening of the planet. As all farmers know, CO_2 is plant food and the emission of increasingly large amounts of CO_2 by humans is good for life on Earth.

ii. The increased carbon dioxide, a greenhouse gas, will lead to ever increasing global warming

Point (ii) has shown to be invalid on all time scales. There is no doubt that CO_2 is a greenhouse gas. However, the main greenhouse gas is water vapour. The first 100 parts per million (ppm) of CO_2 have a significant effect on atmospheric temperature, whereas any increase from the current 400 ppm will have an insignificant effect. Furthermore, because CO_2 has a short residence time in the atmosphere, it is naturally sequestered into the oceans, life, or rocks in less than a decade. In fact, only one molecule of every 85,000 in the atmosphere is CO_2 of human origin, and yet we are asked to believe that this one molecule drives hugely complex climate change systems. We are also asked to believe that the 32 molecules of CO_2 of natural origin in every 85,000 molecules play no part in driving climate change.

Despite a significant increase in industrial emissions of CO_2, there has been no increase in global atmospheric temperature over the past seventeen years. This shows that the theory that CO_2 emissions of human origin drive global warming (and climate change) must be rejected. But this rejection has not yet taken place.

In ice core measurements, the evidence shows that temperature increase occurs hundreds to thousands of years before there is an increase in atmospheric CO_2. This again shows that atmospheric CO_2 does not drive atmospheric temperature change.

On yet another scale, geology shows that all six of the great ice ages were initiated when atmospheric CO_2 was far higher than at present and, with the first two great ice ages, up to a thousand times higher than the current atmospheric CO_2 content. Furthermore, geology shows that there has been sequestration of atmospheric CO_2 into limey sediments, other rocks and life for 2,500 million years. This process continues. The Earth currently has a very low CO_2 content compared with the past. We actually live in a cold epoch. Ice is a rare rock and has been on Earth for less than twenty per cent of its history. For most of time, planet Earth has been warmer and wetter.

In the past, ecosystems thrived when there was a high atmospheric carbon dioxide content—especially if it was warm. This is known by horticulturalists. They pump warm CO_2 into glass houses. The optimum CO_2 content is more than 1,600 ppm (as compared with the current atmospheric content of 400 ppm).

History shows that communities thrived when it was warm and there was massive depopulation during cooler times. It is only recently—when Westerners have been very affluent—that people have feared the warmth. In former times, communities feared the cold because Jack Frost brought death, disease, famine, and war.

iii. There will be tipping points, sea level rise, extinctions and ocean acidification

Sea level changes

Point (iii) is not reflected by evidence. In the past, when atmospheric CO_2 was up to one thousand times higher than at present, there were no tipping points, no carbon dioxide-driven climate change, and no runaway global warming. In fact, the planet was its normal wet-warm self, with the occasional ice age. Sea level rise is caused by water covering the land or the land sinking. Water can rise over the land because the oceans fill with sediment, large submarine igneous provinces displace water, the ocean floor rises, and continental glaciers and ice sheets melt.

The most common reason for water covering the land is ice melting. However, in each of the six major ice ages there were hundreds of glaciations and warm interglacials. Ice retreats and expands for many reasons and temperature is only one of the reasons.

When glaciation locks up more water as snow and ice, sea level falls and the land covered by ice sinks, creating a land rise elsewhere. During an interglacial, sea level rises, the land that was covered with ice rises and other land areas sink. Within the current interglacial, sea level has risen about 130 metres over the last 12,000 years, the rate of sea level rise has decreased (as would be expected towards the end of an interglacial), some land areas that were covered by ice (e.g. Scandinavia) have risen, and near shore ice sheets have been destabilised by sea level rise. Sea level changes are natural. Since the zenith of the last glaciation 20,000 years ago, sea level has risen. What is expected after a glaciation? A sea level fall or a sea level rise? What is important is that the post-glacial rate of sea level rise is declining, exactly what would be expected at the end of an interglacial period. *Nature Geoscience* recently reported that since 2002, the rate of sea level rise has declined by 31 per cent.

It seems illogical that the current sea level rise is due to human activities, whereas the previous hundreds of sea level rises were not. In fact, sea level rises and falls are used by petroleum geologists to understand the 3D shape of potential oil and gas reservoirs. Geologists have known about sea level rises and falls and climate change for hundreds of years, and the community has now only just woken up to the fact that the planet is dynamic. In what can only be regarded as religious narcissism, climate catastrophists now claim that sea level rise, ice sheet melting, torrential rains, drought, hurricanes, and any other severe weather event is due to the activity of affluent Westerners. The past shows a very different story.

Land level changes

There can be no understanding of sea level rise and fall without an understanding of local land level rises and falls. Scandinavia, Scotland and Canada are rising because, during the last glaciation, ice sheets covered these areas and pushed down the land. Now that the ice has melted, there is rebound and the land is rising. If land rises, other areas of land may sink, such as Holland. Land rises in mountains as a result of compression (e.g. Himalayas) whereas, when there is extension or pulling apart, land sinks (e.g. Lake Eyre). The world's oceans formed by extension and, because the oceans are still growing at the mid-ocean ridges, the land masses at the edges of oceans are uplifted into hills or mountain chains (e.g. Great Dividing Range).

These changes are very rapid. For example, the ancient port of Ephesus in Western Anatolia is now fifteen kilometres inland and seven metres above sea level. The ancient Lycian city of Simena on the southern coast of Anatolia is now underwater. Coastal areas may sink due to fluids such as water, gas and oil being expelled from unconsolidated sediments during sediment loading, traffic, buildings, human-induced vibrations, and tides. Any local government that brings in legislation to restrict coastal building by using international sea level projections has only used half the evidence, albeit questionable anyway, as local land level rises and falls are far more profound that long-term sea level rises and falls. The Maldives is 70 centimetres higher now than in the 1970s and eastern Australia is two metres higher than 4,000 years ago. Without a detailed knowledge of local land rises and falls, subsidence, erosion and sedimentation, global sea level predictions for coastal planning are only unfounded speculation.

Coral atolls

Charles Darwin showed in 1842 that as sea level rises, coral atolls grow and keep up with the sea level rise. His suggestion was that coral atolls growing on top of a volcano keep growing at a very rapid rate as the volcano subsides. The sinking of an island has the same effect as a sea level rise. It is a relative sea level rise. Darwin's theory was validated after drilling of coral atolls in the South Pacific Ocean in the late nineteenth and mid-twentieth centuries. His theory was again validated by drilling coral atolls in the Bahamas.

Elsewhere in the Pacific (e.g. Vanuatu), a local land level rise has elevated coral reefs above sea level and dead modern coral reefs occur well above sea level. If Pacific island nation states enjoyed a sea level rise, their land area would increase. This was suggested by Darwin and has been confirmed by recent satellite measurements. Compaction, use of coral for cement manufacture, roads and construction, and extraction of ground water from unconsolidated coralline sand all lead to a relative sea level rise in the Pacific islands, as does polar ice cap melting.

Extinctions

Extinction is normal. Highly adapted terrestrial species (such as humans) have a short life whereas some basic highly adaptable species can survive for billions of years (e.g. bacteria). There is a great diversity of reasons for extinction and climate change is only one of the minor causes. At any time in the history of the planet, there are extinctions, and hence it is no surprise that we live in a period of extinction. Dominant species, such as humans, kill other species for food and change habitats. It has happened many times before. Vacated ecologies are quickly filled and life goes on.

Global warming may create a few extinctions although most species (including plants) have the ability to move to their ideal climate. The history of the planet shows that there is a huge increase in biodiversity during warm times and that extinctions are universal in colder times, when ecosystems are reduced or changed. For example, during the last glaciation, the Amazonian rainforests did not exist and there were copses of trees and grasslands. Inland Australia was devegetated and covered by shifting sand dunes.

The Australian Great Barrier Reef, the poster child of the Greens, disappeared during glacial events more than 60 times over the last three million years. It reappeared after every one of these events. The Great Barrier Reef first formed about 50 million years ago and has survived hundreds of coolings and warmings and massive rain events that deposit sediment on the Reef. The sea levels fall and lower temperatures during glacial events kill higher latitude coral reefs and they continue to thrive at lower latitudes. The geological record shows that coral reefs love it warm, especially when there is more CO_2 in the atmosphere. During glacial events, tropical vegetation is reduced from rainforest to grasslands with copses of trees, somewhat similar to the modern dry tropics inland from the Great Barrier Reef.

Ocean alkalinity

In former times of high atmospheric CO_2, oceans were not acid, there was no runaway greenhouse, and the rate of change of temperature, sea level and ice waxing and waning was no different from the present. The alkalinity (measured as pH units on a logarithmic scale) of ocean water changes is slightly variable. A very slight change to ocean pH would involve a chemical reaction utilising monstrous volumes of acid. Seawater does not become acidified: it changes slightly in alkalinity. The lowest alkalinity (pH 7.3) is very close to acid hot springs. Any Green, climate activist, or journalist who refers to ocean acidity demonstrates a lack of knowledge of basic chemistry. Or maybe they are just deliberately misleading.

The oceans have been alkaline throughout the history of time because water chemistry, ocean floor sediments, and new volcanic rocks on the sea floor buffer seawater to stop it becoming acid, even during times of CO_2 concentrations that were thousands of times the present value. Ocean waters, such as borates, buffer seawater and keep its pH constant. At mid-ocean ridges where volcanic rocks spew out on the ocean floor above large magma chambers, the extensional tectonics allows the ingress of cool alkaline seawater down fractures to depths of about five kilometres into the fresh basalts.

Chemical reactions between natural glasses and minerals in basalt cause water and rock to swap chemicals. This is a buffering process that allows the oceans to remain at constant pH. This process has been taking place for thousands of millions of years during warm times, cold times and times of high atmospheric CO_2 yet the oceans have never been acid. If they were acid at some time, then there would have been a gap in the marine fossil record as carbonate shells of organisms would have dissolved. There is no such gap.

iv. Climate change will be irreversible and that human emissions of carbon dioxide must be reduced or stopped as soon as possible

Governments and their agencies claim that science supports their ideology, but while research grants are given to support this ideology, naysayers are denied grants, ignored, or—more commonly—pilloried. This doesn't happen in many other branches of science, where competing theories are supported with research funds, ideas are energetically discussed, and theories are changed based on new validated evidence. Matters of climate change have been politicised, everyone has an opinion (despite commonly not having the knowledge to underpin an opinion), scientifically illiterate journalists become champions of a cause rather than impartial journalists, and various media networks have taken a partisan political position.

There has never been a public debate about human-induced climate change. Only dogma. Science is full of different interpretations of similar observations and, while it sometimes leads to heated and protracted arguments, it seldom leads to one side trying to attribute to their opponents all the basest characteristics of the human species. Yet this is precisely what happens in the climate change non-debate. Question even one minor factor in the 'official' story and you are likely to be accused of all sorts of political chicanery and moral turpitude. I am yet to find a scientist or read a paper which claims that the climate is not changing. Hence, to label someone as a climate change 'denier' demonstrates that the accuser believes that without human activity, climate would not change. This is ignorance.

If Australia reduced its CO_2 emissions by 5 per cent by 2020, unvalidated models by climate 'scientists' predict that there would be a cooling of between 0.0007°C and 0.00007°C. Such temperature changes are experienced by just moving. This temperature decrease cannot be measured, and such a restriction of emissions is pointless in the light of the great increase in emissions by the developing world. Surely, few activists would consider this meaningful. Australia would suffer an alarming fall in its standard of living and the voluntary act of international environmental kindness would have absolutely no effect on the global climate. Such a self-destructive sacrifice by Australia would not be reciprocated by developing nations such as China and India.

The community sits back with a warm glow feeling that by taxing the 'polluter', it has done something for the planet. They certainly have done something for the planet. The economically vulnerable have been pushed into fuel poverty. Vulnerable people die earlier, costs and unemployment increase and, in the Third World, such climate policies create the continuation of crippling poverty and unnecessary deaths, especially amongst women and children.

In the UK over the last five years, home heating costs have risen 63 per cent, real wages have decreased and an increasing number of the poor spend more than ten per cent of their income on energy. Energy poor pensioners are spending their days riding in heated buses to keep warm, a third are leaving parts of their homes cold and rugging up with hats and scarves and blankets and they are forced to stay in bed longer because of the cost of energy. Is this green policy about helping the poor or the triumph of ideology over tried-and-proven systems?

This warm embrace of feel-good, highly expensive, 'renewable' wind energy has left the most vulnerable citizens out in the cold. Literally. In Germany, charities report the power is cut off from more than 300,000 households each year because consumers can't afford to pay the high costs of 'renewable' green electricity. Some 800,000 Germans are now described as being in energy poverty. German consumers now will be forced to pay annually more than €24 billion to subsidise electricity from solar, wind and bio fuel generating plants that produced electricity at a market price of just over €3 billion. Because of the green dream, Germans now have the highest electricity prices in Europe.

In the UK, green levies for 'renewable' energy are causing energy poverty for 2.4 million British households. There are some 6,000 wind turbines there, with about 1,000 offshore. In the 2012-2013 winter, there were 35,000 additional deaths. This correlates with the increase in wind turbines and the increasing number of the people facing energy poverty. It translates as six elderly, sick or vulnerable people dying each year for every wind turbine, or six deaths per megawatt of wind power generated.

In the 2011-2012 winter, tens of thousands of trees disappeared from parks and woodlands across Greece. Impoverished residents did not have money to pay for electricity and turned to fireplaces and wood stoves for cooking and heat. The same has occurred in Germany. The combination of a cold winter and rising energy costs has forced people to go collecting wood in the forests for home heating and cooking.

Governments cannot resist a new tax. The community sees carbon taxes as a tax on rich, filthy industrialists and their polluting businesses. There is no such thing as carbon pollution and the carbon gas emitted from industry is CO_2—a colourless, odourless, non-toxic gas. CO_2 is plant food. It is good for life on Earth, and human emissions are directly proportional to employment. There is a very low level of the community scientific knowledge displayed when CO_2 is regarded as a pollutant rather than the key to photosynthesis.

v. In order to stop climate change, energy sources need to be shifted from coal, gas and oil to wind, solar, tidal and biomass

The 'alternative' energy systems such as wind and solar are environmentally disastrous. They cause loss of ecosystems, destruction of wildlife, sterilisation of land, inordinate costs that may not be retrieved during the life of the system, and the emission of huge amounts of CO_2 during construction. Furthermore, both wind and solar power are inefficient. They can't provide 24/7 base-load power and need backup by coal-burning carbon dioxide-emitting electricity generating plants. If Australia were to generate 50 per cent of its energy needs from wind, an area the size of Tasmania would have to be clear-felled and covered with wind turbines, because wind energy is low density. One large nuclear- or coal-fired power station occupying a few hectares would generate the same amount of energy.

Germany shut down eight nuclear power stations because of Green pressure. Although Germany has a huge solar and wind generating industry, the unreliability of these ideological power sources is such that Germany has now increased its CO_2 emissions by building new thermal coal-power stations. German electricity prices are now almost twice those of the US, and it is hurting. Ironically, the coal boom in Germany is a result of Greenpeace's political success.

Denmark had been a very enthusiastic supporter of wind energy, but in 2004, it decided to build no more wind farms because it was producing the most expensive energy in Europe. Denmark could see the financial writing on the wall. Although the Danes had become dependent upon wind energy, they found that when the wind did not blow they could not buy wind-generated electricity from north Germany because the weather conditions were the same there. They resorted to buying more reliable hydro- and nuclear-generated electricity from Norway or nuclear-generated electricity from France at high prices. When the wind was strong, the power could not be sold because it was also strong in north Germany. This electricity had to be given away. Denmark now has green taxes that account for more than 50 per cent of an electricity bill.

The US Environmental Protection Agency has engaged in a campaign essentially to regulate coal-fired electricity generation out of existence in the USA. 29 US states and the District of Columbia now have 'renewable' energy mandates and many are trying to impose cap-and-trade programs. If indeed humans are changing climate, funds that could be dedicated to helping people prepare for and adapt to climate change and extreme weather events are wasted on futile attempts to stop what might (or might not) possibly happen in 50 or 100 years. The US alone spends $7 billion each year on 'warming studies' which, in truth, is nothing but a huge money laundering operation, since no real science is conducted. Vapid alarmist reports are the only product generated.

Electricity from the wind is totally unreliable, uneconomic, and degrades the environment. Wind energy neither decreases CO_2 emissions nor changes global climate. No wind farm could operate without generous taxpayer subsidies and increased electricity charges to consumers and employers. These subsidies are given irrespective of whether the wind farm produces any consumable energy or not and are paid even when a wind farm is shut down due to strong winds. Wind farmers have been more successful in harvesting massive subsidies from taxpayers than harvesting the wind. The subsidies in Australia are paid per megawatt generated via a 'renewable' energy certificate. More bureaucratic jobs are needed.

Wind farms produce less than 30 per cent of their nameplate capacity, often at times of low electricity demand and low electricity prices. No carbon dioxide-emitting coal-fired thermal power station has been replaced by a wind farm. Reliable, tried and proven, low-cost, efficient electricity generation from coal is needed as backup because most of the time the wind does not blow or blows too strongly. Coal-fired power stations take 24 hours to fire up and they just can't be turned off and on depending upon whether the wind decides to blow or not. In still cold weather, wind farms consume electricity from coal-fired power stations to stop lubricants freezing.

Industrial economies and urban areas need low-cost efficient electricity to function. Eventually, subsidies will run out and the countryside that was once beautiful will be left with defunct wind farms as a memorial to arrogant green stupidity. In many places, there is no bond held for decommissioning wind farms and land rehabilitation, hence we will be left with memorials for our energy stupidity for generations.

Wind farms have the lives of parasites. They cannot produce continuous electricity without coal, gas, nuclear, hydro or geothermal backup. They freeload by attaching themselves to an existing electricity grid built and paid for by those using conventional energy.

Each January-February, the Northern Hemisphere has a cold snap and wind just does not blow. People die. In southeastern Australia in January 2014, the grid needed 12,000 megawatts at peak when the temperature was more than 40°C for days. The 28 wind farms in southeastern Australia could only provide 128 of the 12,000 megawatts required and it was coal that provided the electricity for air conditioning. When wind farms were needed to provide much needed electricity for cooling, they only operated at less than 5 per cent capacity.

Furthermore, during the 45°C heatwave on 14 January 2014, South Australian electricity wholesale prices spiked at $10,515 per megawatt hour. Many people were without electricity. This dwarfed the wholesale long-term spot price of $70 per megawatt hour. In South Australia, 40 per cent of the electricity is supposed to come from wind power. It doesn't.

If the wind were constantly blowing at eleven metres per second at every wind farm in South Australia spread over hundreds of kilometres, then the nameplate capacity of 1,203 megawatts would be generated.

This does not happen. The Greens state that the wind is always blowing somewhere over such an extensive area so power is always being produced. Reality is different and this does not happen.

In reality, only 60 per cent of South Australia's notional generating capacity is available to service demand when wind watts go walkabout over 100 times a year. When there is no wind, open cycle gas turbines (at $300 per megawatt hour) and 65 megawatts of diesel generators at the defunct Adelaide Desalination Plant kick in to generate electricity and make a killing at the expense of the consumer.

Wind turbines all tend to produce peak power at the same time, when winds are strong. They also all produce nothing when there is no wind. This surging creates huge transmission network problems and, at times, the network is over-capacity. Because of this, wind power is very sensitive to wind speed and can only operate at low wind speeds and therefore is the lowest quality power for the grid. At other times, it is under-capacity.

Conclusion

Climate change catastrophism is the biggest scientific fraud that has ever occurred. Much climate 'science' is political ideology dressed up as science. There are times in history when the popular consensus is demonstrably wrong and we live in such a time. Cheap energy is fundamental for employment, living in the modern world, and for bringing the Third World out of poverty.

As a result of noisy minority political pressures, Western democratic governments have increased energy costs and created subsidised energy systems that have created a new source of tax revenue. Politicians have responded to a groundswell of unscientific environmental concerns rather than make hard decisions. The end result is increased unemployment, lack of competitiveness, energy poverty and increased costs. Unless nature has another surprise for us, three short decades of irresponsible climate policy will take at least a generation to reverse because there are now armies of bureaucrats, politicians, scientists, and businesses living off the climate catastrophe scare. Furthermore, the education system has been captured by activists, and the young are inculcated with environmental, political, and economic ideology. During their education, these same young people are not given the basic critical and analytical methods to evaluate ideology that has been presented as fact. Only a brave government can change the education system to one that prepares people for life.

2 Why climate models are failing

Patrick J. Michaels

Onerous and ineffective policies related to greenhouse gas emissions—such as cap-and-trade schemes, taxes on extraction, or carbon taxes—have themselves extracted a very high political cost in Australia. Three national leaders in Australia—Malcolm Turnbull, Kevin Rudd, and Julia Gillard—were all removed, either by their own party or by the voters, because of failed climate change policies. Environmental policies are also critical issues in state elections, where only one state premier remains from the Labor Party.

In the United States, 63 Democratic Party seats were lost in the House of Representatives in the 2010 election. Almost every close race was lost by a member who voted for cap-and-trade. In the Senate, which did not pass or even bring up similar legislation, every close race went to the Democratic candidate. Given that both Houses had voted for the President's unpopular health care nationalisation, the blame for the loss in the House of Representatives lies squarely with cap-and-trade.

In Canada, a newly-elected and popular government has basically shut down any significant global warming policies.

Despite all of this, which is surely known to the political class, the current Australian national government seems committed to yet another round of global warming-related policy.

This chapter will demonstrate that such a dangerous political course is simply no longer justified by climate science (indeed, if it ever was). It is hoped that those considering new policies may be moved by its contents.

Scientific introduction

In its most basic form, science consists of statements of hypotheses that are retained by critical tests against observations. Without such testing, or without a testable hypothesis, Karl Popper stated that what may be called 'science' is, in fact, 'pseudo-science.'[1] A corollary is that a theory which purports to explain everything in its universe of subject matter is, in fact, untestable and therefore is pseudo-science. In climate, perhaps it is charitable to refer to untested (or untestable) climate model projections as 'climate studies' rather than 'climate science.'

The popular literature is replete with stories relating all kinds of mayhem to climate changes induced by increasing the infrared absorption of the lower atmosphere, known colloquially (and erroneously) as the 'greenhouse effect'. The chain of causation usually cited goes back to one of the very large number of general circulation climate models (GCMs) now in existence.

Given that there are—at least in theory—only two major anthropo-generated alterations in the atmosphere used in climate forecasts, namely an increase in infrared absorption in the lower atmosphere (from CO_2 and other greenhouse gases) and an increased backscattering of incoming radiation (from anthropogenerated aerosols), all subsidiary effects must derive from these two phenomena.

There is one variable in GCMs that changes all the subsidiary phenomena, such as cloudiness, lapse-rate derived precipitation, or changing intensity and distribution of the dynamic systems that initiate rainfall and snowfall. *It is an average warming of the lower atmosphere and a related change in the vertical lapse rate.*

A peculiar phenomenon in modern climate studies is that cause and effect are not easy to parse, and in some instances they are not even explored. In a prominent recent example, President Obama's science advisor, John Holdren, citing the work of Francis and Vavrus[2], suggested that the cold and snow of the winter of 2013-14 in the eastern United States was caused by a warming-related modulation of the circumpolar vortex fueled by declining Arctic sea ice. This was subsequently shown to enjoy no support in the observational data by Barnes and Ballinger et al.[3] (Holdren nonetheless persisted with this meme in support of President Obama's 'Climate Action Plan.')

What is truly odd is that there is no comprehensive examination in the refereed literature of the behavior of the entire community of climate models, when it comes to their 'prime mover' output, a warming of global average surface temperature. Our interest in this has been piqued by 'the pause' in warming, in which annual data in the scientifically popular HadCRUT4 temperature history of Morice et al.[4] are now in their eighteenth consecutive year without a statistically significant warming trend in global average surface temperature.('HadCRUT ' is the name given to the combination of the data sets of two organisations namely the Hadley Centre of the UK Met Office and the Climatic Research Unit (CRU) at the University of East Anglia).

Additionally, over the last thirty years (1984-2013) the suite of 108 climate model runs used in the 2013 compendium of the Intergovernmental Panel on Climate Change (IPCC) produces an average surface warming rate of 2.6°C per century, while the observed value was 1.7°C, a considerable difference for so long a period.

Materials and methods

We examined data since 1950 for the 108 model runs used in the Working Group I (Science) 2013 IPCC *Fifth Scientific Assessment* available from KNMI Climate Explorer (climexp.KNMI.nl). We calculated the model trends for periods beginning at ten years (i.e. 2004-2013), eleven years

(2003-2013), etc., all the way back to 1951-2013.

For each trend length, we ranked the 108 trend values from the individual model runs. From this ranked data set, we determined percentiles. Given the sample size, directly obtaining percentile rankings better characterises and constrains the properties of the data than probabilities derived from the assumption of normality. Inspection shows that the data are not grossly non-normal as do other similar analyses examining the collective trends from climate model projections.[5]

We calculated the average warming produced by the 108 models, as well as the following percentile values: 2.5, 5, 95, and 97.5. The fifth and 95th percentiles are given as the light grey lines in Figure 1, while the broken lines are the 2.5[th] and 97.5[th] percentiles. The average is given by the solid grey line near the midpoint of the percentile bounds. We then compared these trends to the observed trends in the HadCRUT4 temperature history.[6] These are given as the large circles.

Figure 1: Observed vs modeled surface temperature trends (trends end in 2013)

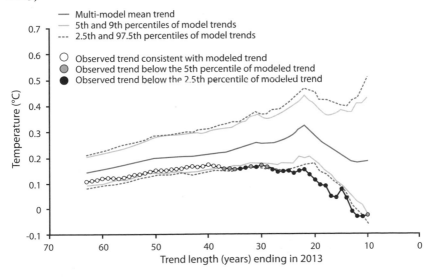

Source: P.J. Michaels

Where they are not shaded, the observed trends are within the 95 per cent confidence limits of the model ensemble. Where they are light grey, the observed trends are below the fifth percentile of the ensemble, and where they are black, they are below the 2.5th percentile.

The results

The results are completely shattering for the ensemble behavior of the 108 model runs. Several items are worthy of note:

- Every observed trend, from 1951-2013 to 2004-2013 falls below the model average.
- The observed trend initially falls below the fifth percentile trend 37 years ago, or in 1977, and remains there for every trend length through the end of the record.
- The observed trend initially falls below the 2.5th percentile trend 34 years ago, or 1980.
- Since 1980, there are only four trends between the 2.5th and fifth percentiles.

If policies were based upon climate science rather than climate studies, this simple, straightforward analysis would spell the end of any onerous climate policy. However, while our similar studies can be scientifically cited,[7] to date, there has been an understandable reluctance to publish this in the tier-1 scientific literature, such as *Nature* or *Science*, as that would indicate a massive, unexplainable, and persistent failure of the studies driving global climate policy.

Paltridge recently speculated that when this is ultimately permitted, the cost to all science (not just climate science) is going to be dear and lasting, much to the detriment of our society and our public policies.[8] It will provoke serious doubt that the present incentive structure in science—which requires that the practitioners keep their problems 'important'—has far-reaching and disastrous unintended consequences.

Kuhn attempted to explain the reluctance of a scientific community to abandon a failing paradigm as a function of the profession, in which the vast majority of practitioners advance professionally by trying to explain minutiae or anomalies within the paradigm.[9]

Therefore:

> In science ... novelty emerges only with difficulty, manifested by resistance, against a background provided by expectation. Initially, only the anticipated and usual are experienced even under circumstances where the anomaly is later to be observed.[10]

It is noteworthy that Kuhn first wrote his manuscript in the late 1940s, which was prior to the completion of the large-scale transition of science to essentially a publicly-funded enterprise. Consequently, he does not explore how the need to keep public funds flowing through academia probably made paradigms more 'sticky' than they already are.

Additionally, Kuhn was encouraged in his work by his mentor, James Conant, Harvard's president and a major figure in the Manhattan Project, the largest public science venture (at that time) in history. The Manhattan Project and the Office of Scientific Research and Development were the prototypes cited by President Roosevelt when he requested the federalisation of science in a 1944 letter to its director, Vannevar Bush.

It is therefore hard to imagine that Kuhn had much of an incentive to note how massive public funding might have a less-than-salutary influence upon scientific progress. That was certainly not in the interests of his mentor or his employer.

A problem that deserves much further investigation is how climate science could continue in its remarkable denial that the aggressive global warming paradigm has been shattered, with now 37 consecutive years of documented, systematic model failure.

Is an alternative paradigm developing?

Beginning in 2011, an increasing number of papers and modelling experiments began to appear in the literature indicating, either from models or from historical observations, that the sensitivity of the paradigmatic family of GCMs is too high. 'Sensitivity' is the ultimate amount of surface warming that is realised for a doubling of the ambient CO_2 content from its preindustrial background. This is nominally construed as a change from 300 to 600 parts per million (by volume).

Lindzen and Choi used discrete warming and cooling periods of the tropical ocean along with outgoing radiation at the top of the atmosphere as measured by two satellites and concluded that the resultant temperature feedbacks to thermal changes in the ocean and outgoing radiation implied a lower sensitivity of temperature to longwave ('greenhouse') radiative forcing, perhaps because of negative feedbacks within the earth-atmosphere system.[11] According to Lindzen and Choi, observational data 'imply that the models are exaggerating climate sensitivity.'[12]

Schmittner et al. combined large-scale temperature reconstructions from the last glacial maximum with various model simulations (which include changing CO_2) and calculated a mean sensitivity of 2.3°C compared to the IPCC Fifth Assessment average of 3.2°C .[13] More importantly, the probabilities of very large warmings are dramatically reduced. They conclude that '[a]ssuming paleoclimatic constraints apply to the future as predicted by our model, these results imply lower probability of imminent extreme climatic change than previously thought.'[14]

While noting that the sensitivity is far from certain, Annan and Hargreaves used probabilistic estimates themselves rather than a 'uniform prior' subject to perturbation by multiple GCM runs, noting that the standard methodology 'has unacceptable properties' that condition it to produce unrealistically high probabilities for large warming.[15] By using the more realistic approach, the 95 per cent confidence limit for a large warming dropped by one-third, and they noted that '[t]hese results also impact strongly on projected economic losses due to climate change.'[16] Incorporating a prior probability distribution based on expert opinion (rather than the oft-used and ill-founded uniform distribution) into a Bayesian analysis, Annan and Hargreaves find a median value for the equilibrium climate sensitivity of 1.9°C. Incorporating a Cauchy distribution for the Bayesian prior results in a median climate sensitivity estimate of about 2.2°C.

Van Hatteren found, using a retrospective approach, a climate sensitivity of 2.0 +/- 0.3°C, and noted 'it is at the lower end of the range considered likely [by the IPCC].'[17] In this they were referring to the 2007 IPCC 'Fourth Assessment' report. The 2013 report reduces the lower limit from 2.0 to 1.5°C, with no 'best estimate'. This will provide the IPCC with some cover when the low sensitivity becomes obvious even to those currently defending the established, failing paradigm.

One of the most important papers in this tranche is that of Ring et al.,[18] if only because the fourth author, Michael Schlesinger of the University of Illinois, has long been an extremely vocal advocate for a high-end sensitivity. By adjusting their model with observed temperatures (which includes much of 'the pause'), they arrive at a sensitivity of 1.5-2.0°C. While they claim that this is 'on the low end of the estimates in the [2007] IPCC's *Fourth Assessment Report*' it is in fact clearly beneath that[19]—yet another example of Kuhn's observation that what turns out to be obvious is initially ignored.

Hargreaves et al. used a new determination of the cooling during the Last Glacial Maximum to derive estimates of the climate sensitivity.[20] Their estimates use two different statistical techniques, one employing regression relationships between tropical temperatures during the Last Glacial Maximum and climate model climate sensitivity, and another using a Bayesian approach weighting each climate model based on how well it matches the new Last Glacial Maximum data. The two methods produced very similar results, with a mean equilibrium climate sensitivity of about 2.5°C with a 90 per cent confidence range of about 1°C to 4°C.

Aldrin et al. also fit model results to observed temperatures but in addition included observed changes in oceanic heat content and found a mean sensitivity of 2.0°C, which is over 40 per cent below the mean of the model family used in the 2013 IPCC report.[21]

Spencer and Braswell used a simple climate model to simulate the global temperature variations averaged over the top 2000 meters of the global ocean during the period 1955-2011.[22] They first ran the simulation using only volcanic and anthropogenic influences on the climate. They ran the simulation again adding natural variability contributed by the El Niño/La Niña process. Then they ran the simulation a final time adding in a more complex situation involving a feedback from El Niño/La Niña onto natural cloud characteristics. They then compared their model results with the set of real-world observations. They found that the complex situation involving El Niño/La Niña feedbacks onto cloud properties produced the best match to the observations, and, notably, produced the lowest estimate for the earth's climate sensitivity to CO_2 emissions—a value of 1.3°C. Climate models which do not accurately recognise this feedback will produce climate sensitivity values from the doubling of the atmospheric concentration of greenhouse gases far in excess of this observationally-determined value.

Otto et al. is highly significant because fifteen of the seventeen co-authors were also authors of the 2013 IPCC report.[23] They acknowledge both 'the pause' and recent heat uptake and note that these yield a lower sensitivity than has been found in previous models. They prefer using the most recent decade because it contains more reliable baseline energy balance data, and they calculate a five to 95 per cent confidence range of 1.2-3.9°C, and a mean sensitivity that is also more than 40 per cent less than the average in the 2013 IPCC report. Why this was not also a conclusion of that report, given the position of so many of its authors, seems mysterious.

Lewis, in a highly-cited paper, tuned a climate model to provide an 'optimal fingerprint' to calculate the temperature sensitivity most consistent with the observed diffusion of heat through the ocean and changes in atmospheric transparency.[24] Their 90 per cent confidence interval is 1.0-3.0°C, virtually the same range found by Aldrin et al, noted earlier.[25] Lewis notes he used the same basic temperature statistics as did Aldrin.

Masters, using the observed changes in heat content of the upper 2 kilometers of the ocean, along with temperature observations since 1950, argued that the model family employed in the 2013 IPCC report is too sensitive to changes from both warming greenhouse gases and cooling aerosol emissions.[26] Their results suggest that the mean equilibrium climate sensitivity is around 2.2°C, while the average of the IPCC models is 3.2°C.

Most recently, Loehle recognized that calculations relating human-induced changes in the energy balance with observed temperatures also yields the now-familiar 2°C sensitivity, but that the temperature trends are also modulated by long-term natural oscillations, such as the Pacific Decadal Oscillation and the Atlantic Multidecadal Oscillation.[27] Removing these from the surface record, and comparing them in a very straightforward relationship to anthropogenerated emissions yields the same sensitivity but also increases confidence in that value, leading him to conclude 'that higher estimates derived from climate models are incorrect because they disagree with empirical estimates.'

Figure 2 graphically summarises the various results in these papers, as well as the mean value and range of the family of climate models used in the 2014 IPCC report. It also summarises the 'likely' range of warming given in that report. See our figure caption for details.

Andrew Revkin, who follows global warming science for *The New York Times*, published that influential newspaper's first article on the emerging evidence for a lower climate sensitivity in 2013.[28] In it, he noted there was one study, ten years earlier, that also projected warming that is consistent with the recent lower-sensitivity results. However, Revkin chose not to specifically identify the authors, saying instead:

> I can understand why some climate campaigners, writers and scientists don't want to focus on any science hinting there might be a bit more time to make this profound energy transition. (There's also reluctance, I'm

Figure 2: Climate sensitivity estimates from new research beginning in 2011 compared with the assessed range given in the IPCC Fifth Assessment Report (AR5) and the collection of climate models used in the IPCC AR5

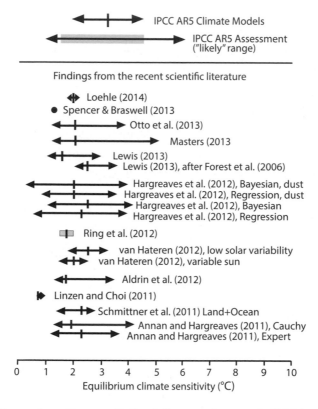

The 'likely' (greater than a 66 per cent likelihood of occurrence) range in the IPCC Assessment is indicated by the gray bar. The arrows indicate the five to 95 per cent confidence bounds for each estimate along with the best estimate (median of each probability density function; or the mean of multiple estimates; vertical line). Ring et al. present four estimates of the climate sensitivity and the box encompasses those estimates.[29] The right-hand side of the IPCC AR5 range is actually the 90 per cent upper bound (the IPCC does not actually state the value for the upper 95 per cent confidence bound of their estimate). Spencer and Braswell produced a single sensitivity value that best-matched ocean heat content observations and the internal ('natural') oscillations they observed.[30]

Source: P.J. Michaels and P. C. Knappenberger, "More Evidence for a Low Climate Sensitivity", available at http://www.cato.org

sure, because recent work is trending toward the published low sensitivity findings from a decade ago from climate scientists best known for their relationships with libertarian groups.)

The unmentioned citation is the Michaels et al. 2002 paper.[31]

Conclusion

There is little doubt that the ensemble of climate models used in the 2013 IPCC report fails, when normatively tested as a hypothesis concerning observed global surface temperature trends. Such a comprehensive analysis has been presented by the authors at scientific meetings with little substantive criticism, but this does not seem to be acknowledged by the larger community undertaking what now must be referred to as 'climate studies', rather than 'climate science'.

The normative test indicates that, when model output from the UN is examined retrospectively, the failure of hypothesis would have begun about 37 years ago. This seems like an inordinately long time for such a problem to go undetected, even stipulating that Kuhn's iconic model of paradigmatic science is applicable.[32]

However, there is a growing body of the refereed scientific literature that estimates, using largely independent techniques, that the sensitivity is lower than the mean in the model ensemble used in 2013 IPCC report. Meta-analyses of these new results have yet to be performed to see whether they, as an ensemble, are also consistent with observed trends.

The failure of the models and the emergence of a new literature need to be acknowledged, as both of these will certainly be invoked in debates about climate change policy. These are particularly timely with regard to the governments of Australia and the United States, both of which have been dramatically changed by the imposition, or attempted imposition of expensive and onerous climate policies.

3   Global warming, models and language
Richard S. Lindzen

A man may take to drink because he feels himself to be a failure, and then fail all the more completely because he drinks. It is rather the same thing that is happening to the English language. It becomes ugly and inaccurate because our thoughts are foolish, but the slovenliness of our language makes it easier for us to have foolish thoughts.

—George Orwell: *Politics and the English Language*

Global warming is about politics and power rather than science. In science, there is an attempt to clarify; in global warming, language is misused in order to confuse and mislead the public.

The misuse of language extends to the use of climate models. Advocates of policies allegedly addressing global warming use models not to predict but rather to justify the claim that catastrophe is possible. As they understand, proving something to be impossible is itself almost impossible.

In a further abuse of language, the advocates attempt to rephrase issues in the form of yes-no questions:

- Does climate change?
- Is carbon dioxide (CO_2) a greenhouse gas?

- Does adding greenhouse gas cause warming?
- Can man's activities cause increases in greenhouse gases?

These *yes-no* questions are meaningless when it comes to global warming alarm since affirmative answers are still completely consistent with there being no problem whatsoever; crucial to the scientific method are 'how much' questions. This is certainly the case for the above questions, where even most sceptics of alarm (including me) will answer yes.

To a certain extent, therefore, this issue cannot be discussed between opponents. We are speaking different languages.

That said, it should be recognised that the basis for a climate that is highly sensitive to added greenhouse gases is solely due to the behavior of the computer models. Within these models, the primary effect of increases in greenhouse gases is multiplied several fold by the interaction of the increases with water vapour, clouds, and other aspects of the system that are openly acknowledged by the Intergovernmental Panel on Climate Change (IPCC) to be highly uncertain. The relation of this sensitivity to catastrophe, moreover, does not even emerge from the models, but rather from the fervid imagination of climate activists.

What are some questions that are relevant?

1. What is the sensitivity of global mean temperature to increases in greenhouse gases?
2. What connection, if any, is there between weather events and global mean temperature anomaly?
3. Is the notion of global mean radiative imbalance driving global mean temperature relevant to actual climate change? The meaning of this question will become evident below.

The above hardly exhausts the list of relevant questions, but in the present essay, I'll focus on the first item, though brief attention will be given to the remaining two questions.

Climate sensitivity

The term climate sensitivity has come to refer to the equilibrated response of global mean temperature anomaly to a doubling of CO_2. Because of the logarithmic dependence of the radiative impact of CO_2, it doesn't matter what the starting value for the doubling is. Doubling from 1000 to 2000 ppmv (parts per million by volume) has the same effect as doubling from 280-560 ppmv. The very definition of climate sensitivity presupposes the answer to the third question is affirmative. While this may be true for more or less global forcing such as that due to increasing a well mixed gas like CO_2, the paradigm is almost certainly inappropriate for major climate variations in the earth's history.

Estimates based on the temperature record

Using the surface temperature record to estimate sensitivity seems like an obvious approach, but, in fact, it has substantial problems. Nevertheless, as we shall see, the record offers no support for high sensitivity and strongly suggests low sensitivity. Several researchers have attempted to use the record to plausibly establish likely bounds for sensitivity, and these point to a range of lower sensitivities than suggested by model results.[1] However, none of this matters to those arguing for dangerous warming. As long as any possibility that sensitivity is high exists, they are content. There is an implicit (and sometimes explicit) appeal to the precautionary principle. Of course, the precautionary principle, for all its faults and fundamental incoherence, was not meant to suggest that one acts on the basis of possible but highly unlikely dangers. Rather, it suggested that likely, but not rigorously proven, dangers be acted upon. For the advocates of alarm, this merely means that they must insist, against the evidence, that the danger is self-evident—often based on the agreement of almost all scientists with the four trivial yes-no statements at the beginning of this essay.

It is relatively simple to show how such considerations play out in practice using what are called energy balance models. In contrast to General Circulation Models (GCMs) where climate sensitivity is internally generated, these models assume a sensitivity and investigate the response of the surface temperature where heat capacity is that of a relatively simple ocean model. When the chosen sensitivity is that of a GCM, the resulting evolution of temperature follows that of the GCM. As a result, these simple models are used extensively by the IPCC for scenario building. The specific model we will use is that described in Richard Lindzen and Constantine Giannitsis' 1998 paper, 'On the climatic implications of volcanic cooling'.[2] Before going into details, it is worth pointing out that these models describe an important feature of climate: namely, that high sensitivities are associated with long response times. Radiative forcing is a flux of energy (i.e. an energy flow per unit area), and sensitivity is a ratio of temperature change to this flux. High sensitivity means that a small flux eventually produces a large temperature change, but, because the flux is small, the change will take a long time.

Our exercise begins with two sources of forcing common to almost all models. The first is the radiative forcing due to anthropogenic greenhouse gases (CO_2, methane, nitrous oxide, etc.). According to the IPCC, this has grown from about zero in 1850 to about 3 watts per square meter (Wm^{-2}) now. This has resulted from the exponential growth of CO_2. Note that we are already at about 80 per cent of what one expects from a doubling of CO_2 alone. The other is the forcing due to major volcanoes. Figure 1 shows the volcanic forcing estimated by Makiko Sato.[3] The values are comparable to greenhouse forcing, but the volcanoes occur in two clusters with an intervening quiescent period. Such clustering is actually characteristic of random processes.

Figure 2a shows the response of the model for various values of climate sensitivity. The observed warming has been about 0.75°C. This is exceeded by almost any sensitivity in excess of 1°C. However, the calculated responses for

Figure 1: Radiative forcing due to volcanic eruptions since 1850

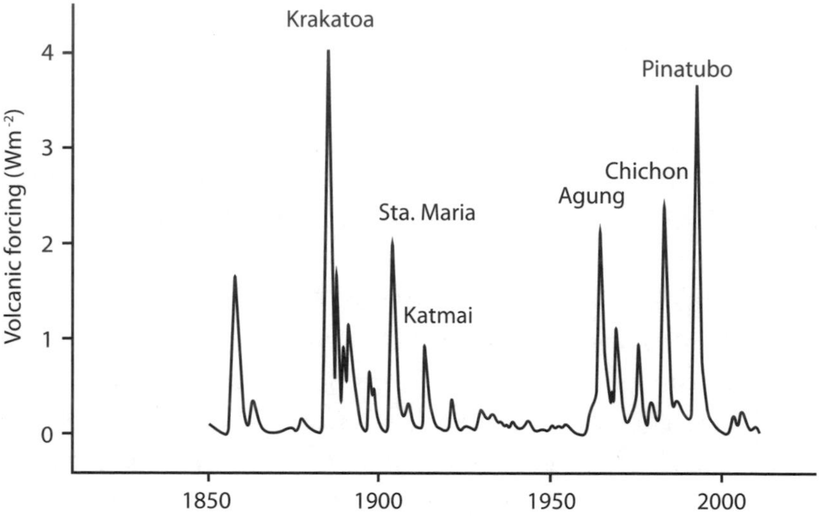

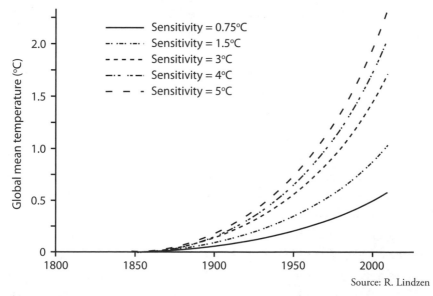

Source: M. Sato, "Forcings in GISS Climate Model," last modified 13 December 2012, http://data.giss.nasa.gov/modelforce/strataer/

Figure 2: Response of global mean temperature to anthropogenic greenhouse forcing for various choices of climate sensitivity

Source: R. Lindzen

models with high sensitivity are smaller than the equilibrium responses because of the long response times associated with high sensitivity.

Figure 3 shows the response of the same models to the volcanic forcing. Here the response time plays a particularly striking role. For low sensitivity, the response consists in episodic dips corresponding to each eruption (much as observed), while for high sensitivities, there is a secular trend leading to a net cooling of about 0.3°C at present. The absence of any evidence of this in the data already points to low sensitivity. However, for the present exercise, this means that the warming due to high sensitivity is moderated to an important extent by the accompanying secular cooling due to volcanoes. Figure 4 shows the response to the sum of greenhouse and volcanic forcing. It is still the case that only sensitivities under about 1°C are consistent with the observations. However, the IPCC suggests that aerosols constitute a highly uncertain potential source of cooling. Modelers have been able to invoke this uncertainty in order to adjust the net anthropogenic forcing (i.e. the sum of greenhouse forcing and aerosol forcing) to match the observations (Figure 5). Table 1 shows how much of the greenhouse forcing has to be cancelled in order to reach agreement. For the sensitivities in excess of 1.5°C, this is about half of the 3 Wm⁻², and already in excess of what the IPCC considers the likely aerosol contribution, but recent work suggests that even this estimate for aerosols is much too great.

Of course, so far, we have assumed that the observed temperature change is entirely due to the combination of greenhouse, aerosol and volcanic forcing. We have ignored the potential forcing due to solar variability, but perhaps more importantly, we have ignored what is called natural internal variability. The fact of the matter is that global mean temperature anomaly would vary even without external forcing. Heat is constantly redistributed by the oceans on time scales ranging from years to millennia, leaving surface temperatures out of equilibrium with radiative forcing.

Figure 3: Response of global mean temperature to volcanic forcing for various choices of climate sensitivity

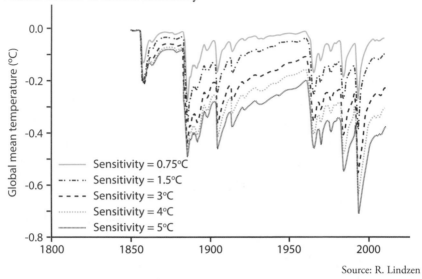

Source: R. Lindzen

Figure 4: Response of global mean temperature to sum of anthropogenic greenhouse and volcanic forcing for various choices of climate sensitivity

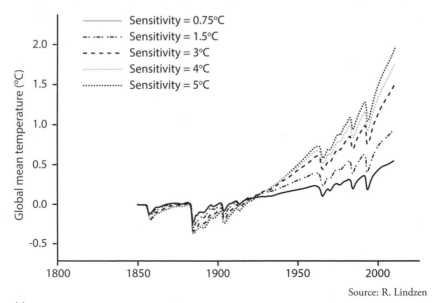

Source: R. Lindzen

Figure 5: Results in Figure 4 'corrected' by aerosols

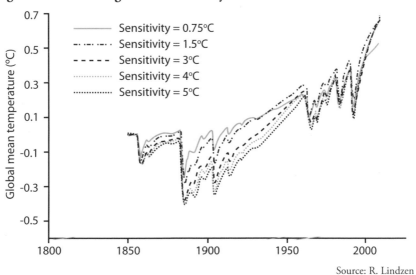

Source: R. Lindzen

Table 1: The amount of 'aerosol' cancellation needed to change results in Figure 4 to those in Figure 5. The latter are consistent with observed net warming

Sensitivity in °C (for doubling of CO_2)	Fraction of greenhouse gas forcing cancelled by 'aerosols'
0.75	0
1.5	0.25
3.0	0.481
4.0	0.525
5.0	0.543

Examples of this natural (non-greenhouse related) internal variability are the El Niño-Southern Oscillation (ENSO), the Pacific Decadal Oscillation and the Atlantic Multi-decadal Oscillation.

The instrumental record is too short to indicate longer period phenomena though such longer periods are evident in proxy records of temperature. The so-called hiatus in global warming over the past seventeen years is incontrovertible evidence of the importance of this variability, and as a 2013 study by Jiansong Zhou and Ka-Kit Tung has shown, such internal variability is likely to have accounted for about half of the warming since the late 1970s.[4] Under the circumstances, getting models to replicate the much reduced directly forced warming will require even more implausible cancellation by aerosols. It should be noted that the IPCC claims for attribution depend on assuming natural internal variability to be small. This is clearly not the case. However, what is generally ignored is that even if the attribution claim were correct, it would still be completely consistent with low sensitivity. After all, the above analysis assumed that more than 100 per cent of the observed warming was due to man (allowing for cancellation of some of the warming by volcanoes). The IPCC only insisted on 51 per cent being due to man.

It seems to me that the most reasonable conclusion to reach under the circumstances is that climate sensitivity is small. Moreover, mild warming is likely to be a net benefit. However, I would not be surprised to see advocates of alarm soon arguing that it was the previously ignored natural internal variability that had actually disguised the large warming that would otherwise have occurred, thus leaving the possibility of dangerous warming in place.

Other approaches to climate sensitivity

It has already been mentioned that high sensitivity is related to long response times and that leads to a pronounced difference in the response to sequences of volcanoes. Figure 6 is the famous figure of the record

Figure 6: Globally averaged temperature anomaly as a function of time

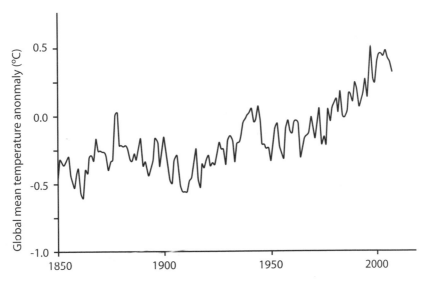

Source: R. Lindzen

of global mean temperature anomaly. Volcanoes appear as isolated dips consistent with low sensitivity (Figure 3). Relatedly, the broad minimum in temperature between 1883 and 1950 seen in Figure 5 is absent in the observed series. Models, however, have openly used hypothetical pictures of aerosols and solar influence to eliminate this discrepancy. Given the numerous possibilities for bringing observations and models into agreement, one might reasonably try to find more direct measurements of sensitivity.

The response time that we have discussed is simply a measure of the thermal coupling of the atmosphere to the ocean. High sensitivity is associated with weak coupling and low sensitivity with strong coupling. Thus, as Gerard Roe noted, it should apply more generally to climate variations other than those simply due to greenhouse forcing.[5] These include the aforementioned El Niño-Southern Oscillation, Pacific Decadal

Oscillation, and Atlantic Multi-decadal Oscillation—all of which depend critically on this coupling and all of which are poorly represented in current models. Seeing whether reducing sensitivity would improve this situation seems like an obvious test, but to the best of my knowledge, this has not been attempted. One is left with the suspicion that the modeling community prefers to defend high sensitivity even when it means failure to simulate known phenomena.

Nir J. Shaviv used correlations between the solar cycle and ocean temperature to estimate the flux needed to produce the change in ocean temperature.[6] This allowed him to estimate the amplification of solar flux (potentially related to cosmic rays and their secondary effects on clouds and albedo), as well as climate sensitivity. Once again sensitivity was around 1°C.

The climate forcing associated with added greenhouse gases manifests itself as radiative forcing at the top of the atmosphere. This forcing also acts on the surface, but no longer in the form of simple radiative forcing. Rather the primary forcing is associated with what is called latent heat flux. This is simply the flux of energy associated with the heat of vaporisation of water vapour that is evaporated from the surface. Here, it has been noted by Frank Wentz et al. that whereas in models evaporation changes about 1-3 per cent per degree of warming, it is observed to change 5.7 per cent per degree of warming.[7] The former is consistent with model sensitivities of about 1.5-4.5°C , but the latter is consistent with much lower sensitivity on the order of 0.8°C.

In the absence of what are referred to as feedbacks, the expected response to a doubling of CO_2 is about 1°C. In current general circulation models, this warming is accompanied by feedbacks to water vapour and clouds (and to a much smaller extent to the reflectivity—known as albedo—of snow) which serve (in the models but not necessarily in nature) to amplify the response to CO_2 alone. Feedbacks that amplify the response are known as positive feedbacks while those that diminish the response are known as negative feedbacks. The situation is crudely described by the following equation (see Roe's 2009 paper for a detailed discussion):[8]

$$\Delta T = \frac{\Delta T_0}{1-\sum f_i} \qquad (1)$$

Where ΔT is the response to a doubling of CO_2, ΔT_0 is the response in the absence of feedbacks, and $\sum f_i$ is the sum of the various feedback factors. A negative f_i clearly diminishes the response while a positive f_i amplifies the response. Should $\sum f_i$ reach +1, the response becomes infinite, but it would take literally forever to reach this value. As we have already noted, high sensitivity is associated with long response times. It is common to consider water vapour and cloud feedbacks separately, but this actually makes little sense. Water vapour is important for infrared radiation, while clouds are important for both infrared and visible radiation. We will refer to the former as longwave radiation and the latter as shortwave radiation. The water vapour feedback is an infrared feedback, but it applies only in regions free of upper level cirrus. The areal coverage of the upper level cirrus (one type of higher level cloud) is highly variable. We must therefore consider the total longwave feedback. Similarly, we can look at the total shortwave feedback.

Direct measurement of feedbacks with satellite data

An obvious approach to measuring feedbacks would be to see how outgoing radiation responds to surface temperature fluctuations. The crucial point about the feedbacks is that they respond to surface temperature fluctuations regardless of the origin of the fluctuations. Not surprisingly, this approach also has difficulties.

Satellite measurements of outgoing radiation (both longwave and reflected shortwave) are used in this approach. It turns out that the model intercomparison program has the models used by the IPCC, forced by actual sea-surface temperature, calculate outgoing radiation. So one can use the same approach with models, while being sure that the models are subject to the same surface temperature fluctuations that applied to the observations.

In principle, this should be a straightforward task. However, as already noted, there are difficulties. The first two difficulties involve basic physical considerations.

First, not all time scales are appropriate for such studies. Greenhouse warming continues until equilibrium is re-established. At equilibrium, there is no longer any radiative imbalance. If one considers time intervals that are long compared to equilibration times, then one will observe changes in temperature without changes in radiative forcing. The inclusion of such long time scales thus biases results inappropriately toward high sensitivity. Equilibration times depend on climate sensitivity. For sensitivity on the order of $0.5°C$ for a doubling of CO_2, it is on the order of years, and for higher sensitivities it is on the order of decades. In order to avoid biasing sensitivity estimates, one should restrict oneself to time intervals less than a year. This problem is particularly acute for approaches which simply regress outgoing radiation on temperature without concern for time scales (viz. the 2006 paper by Piers Forster and Jonathan Gregory, and the 2010 paper by Andrew Dessler).[9] Moreover, given that the response time is shorter for low sensitivity, one has the ironic result that the bias toward high sensitivity is greater when the actual sensitivity is lower.

There is also the need to consider time intervals long enough for the relevant feedback processes to operate. For water vapour and cloud feedbacks, these time scales are typically on the order of days. For practical time resolution, this is generally not a problem.

Time scales on the order of one to three months are, thus, appropriate for sensitivity studies. Longer time scales also involve 'pollution' from seasonal effects, etc. Restricting consideration to such short time scales is the approach taken in papers by Lindzen and Choi, Spencer and Braswell, and Trenberth and Fasullo.[10]

The second problem is more difficult. Outgoing radiation varies (especially in the visible) for reasons other than changing surface temperature (volcanoes, non-feedback cloud fluctuations). Such changes are not responses to surface temperature fluctuations but they do cause surface temperature fluctuations. As the 2014 paper by Yong-Sang Choi et al. shows, the high noise to signal ratio makes determination of the shortwave feedback unreliable.[11] However, the situation is better for the longwave feedback and the results of papers by Lindzen and Choi, Spencer and Braswell and Trenberth and Fasullo all point to either the absence of longwave feedback or negative longwave feedback.[12]

Apart from basic physical issues, there are other practical problems such as the presence of significant gaps in the outgoing radiation data. Also, the radiation data involves two satellite systems (ERBE and CERES) with different properties. The 2011 paper by Lindzen and Choi describes how we deal with these issues.[13]

The fact that the data shows the absence of the longwave feedback is extremely important. In all current GCMs, the contribution of the longwave feedback (generally incorrectly referred to as the water vapour feedback) in the equation above is $f_{longwave} \sim 0.5$. This alone doubles the sensitivity from the no-feedback value. In these models, the contributions of shortwave feedbacks can be as great as 0.3. However, the addition of this to 0.5 leads to an amplification by a factor of five. The variation of the shortwave feedback factor between zero and 0.3 (in models) is what basically leads to the IPCC claim that sensitivity is 2-5°C. However, if the longwave feedback factor is zero or even negative, then the sensitivity with a shortwave feedback factor of 0.3 is no more than 1.4°C and with a shortwave feedback of zero, the sensitivity can well be below 1°C. The absence of a longwave feedback strongly suggests that something like the iris effect discovered in a 2001 paper by Lindzen, Ming-Dah Chou and Arthur Hou is causing the variations in upper level cirrus to cancel any positive water vapour feedback.[14] Consistent with all this is the finding by Lindzen and Choi that no model correctly depicts the observed variations in outgoing radiation.[15]

As we have seen, the basis for the possibility of high sensitivity is either models or clearly incorrect approaches to data. In reality, the data all points to low sensitivity. It is sometimes asserted that paleoclimate data points to high sensitivity. This stems from a profound misunderstanding of the climate system. We will return to this last point later in this essay.

Extreme weather and climate

The failure of the public to get unduly excited over a degree or two of warming has led the environmental alarmists to turn to the bogey man of extreme weather. As even the IPCC acknowledges, there is no empirical evidence for such a connection. Moreover, outside the tropics, the claim runs opposite to what accepted theory implies.

Outside of the tropics, the main process in generating weather disturbances is known as baroclinic instability: an instability whose energy is proportional to the horizontal gradient in temperature. Globally, the strength of this instability depends on the temperature difference between the tropics and high latitudes. In a warmer world, this difference is expected to decrease not increase.

Figure 7 is the temperature map for 11 March 2013 for North America. Air is carried roughly along the path of the jet stream (and this path changes from day to day and year to year). Record breaking temperatures (regardless of the year that they occurred) correspond to the warmest and coldest temperatures on the map for 11 March 2013. In a warmer world, we expect the amplitude of the vacillations in the jet stream as well as its magnitude to decrease suggesting a reduction in extreme events.

Figure 7: Temperature map for North America on 11 March 2013

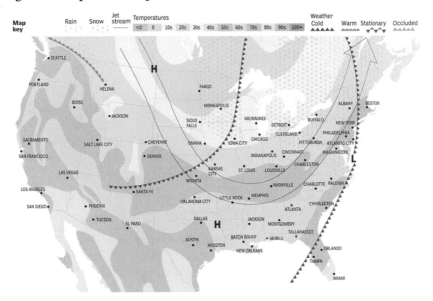

In the tropics, it is frequently claimed that evaporation will increase in a warmer world, and that this will lead to extremes in precipitation, drought, storminess, etc. As already mentioned, there is no statistically significant empirical evidence for any of this, but it is interesting to ask whether even the underlying assertion is true: will evaporation actually increase?

In general, evaporation depends on both the temperature of the wet surface, and the temperature and relative humidity of the air immediately above the surface. The assertion of increased evaporation depends on the assumption that the relative humidity of the surface air in a warmer world will remain unchanged from its earlier value. Not only is there no basis for this claim, but a change in relative humidity from 0.8 to 0.83 is sufficient to completely eliminate any increase in evaporation that would be caused by a warming of 3°C, and such changes in relative humidity at the surface are commonplace and easily produced. The sensitivity to such small changes is not an indication of the delicacy of the climate system, but rather an indication of the ease with which the system can adjust to changes.

In the case of claims of increases in extreme weather, even models do not suggest this to any significant extent. The claims are made without bases for the simple purpose of frightening the public. However, when the public is told that every turn of the weather, whether unusual cold, warmth, snow, drought, etc., is proof of global warming, and when even the term global warming is changed to climate change, the public soon moves the issue into the realm of parody with only the chattering classes continuing to pretend belief in the seriousness of the issue. The writers of insurance policies also find it profitable to take the claims seriously.

Milankovich vs naïve climate picture

We have long known that the earth has undergone ice ages when massive ice sheets covered much of North America and Scandinavia. However, only with the analysis of ice cores from Antarctica have we had a reasonably detailed history of ice volume, temperature, and CO_2. Al Gore's *An Inconvenient Truth* uses such a record to suggest that CO_2 drives climate since the record of CO_2 tracks the temperature record. The CO_2 record shows variations between 180 ppmv and 280 ppmv. The radiative forcing of the high value relative to the low value is about 2 Wm^{-2} which is remarkably small.

Moreover, there is substantial evidence that temperature changes preceded CO_2 changes. During World War I, the Serbian astronomer Milutin Milankovitch argued that orbital changes associated with obliquity cycles, the precession of the equinoxes, and variations in orbital eccentricity should have a profound effect on glaciation. His thinking was eminently reasonable. As is well known, the major factor in long term glacial growth and retreat is summer insolation (the amount of incoming solar radiation). This determines whether snow that accumulates in winter will survive the summer. If it does, then snow will accumulate forming glaciers.

Milankovitch argued that the relevant quantity for forcing glacial cycles is the insolation in summer over the Arctic. The orbital variation of this quantity is about 100 Wm^{-2}, which is huge, compared to the contribution due to CO_2. Moreover, as the 2002 paper by Sverker Edvardsson et al. and the 2006 paper by Roe have shown, the correlation of the Milankovitch parameter with the time rate of change of ice volume is about as good as any correlation in geophysics (earlier claims of poor correlation resulted from using ice volume rather than its time rate of change).[16] Also, the heat flux is consistent with what is needed to produce the phase changes involved in glaciations and deglaciations. On the other hand, the orbital changes result in almost no change whatever in the annually and globally averaged insolation. It is thus suggested that the miniscule radiative forcing due to CO_2 is essential and that the sensitivity is great. What relevance this has to the glaciations cycles is hard to see.

The temperature changes associated with the glaciations cycles are sufficient to account for the changes in CO_2 and the small imbalances are easily accommodated by slight changes in cloud cover, cloud brightness and cloud height—all of which are known to vary substantially more than needed in the present climate.

This suggests that the paradigm of climate represented by a single number (globally and annually averaged temperature anomaly) forced by a globally and annually averaged radiative imbalance is totally inappropriate for the most important climate variability of the last million years.

Conclusion

What we have seen is that the climate is probably insensitive to increases in greenhouse gases, and that there is little reason to suppose that a warmer world will be notably characterised by storminess and extremes though both are part of normal weather variability. Scientific agreement is largely premised on agreement over trivial issues that are distinct from alarm. However, the acquiescence of science in this abuse is disturbing to say the least.

4　Sun shunned

William Soon

Maunder's principal contribution was to emphasize the varying level of solar activity over the centuries since the advent of the telescope, with particular attention to the scarcity of sunspots from about 1645 to 1715, beginning only 35 years after the start of telescopic observations. Maunder's point was conveniently ignored, or even denied, because no one knew what it meant. Fortunately, Jack Eddy took up the historical investigation about 30 years ago and turned up enough old records that the reality of the 'Maunder Minimum' was established beyond any reasonable doubt. Subsequent historical research has unearthed detailed systematic records of sunspot numbers which show how peculiar the behavior of the Sun was during that time. Then with modern data on the atmospheric production rate of carbon 14 by cosmic rays, Eddy went on to show that such prolonged periods of solar inactivity have occurred ten times in the last 7000 years. So we may anticipate that there will be yet another Maunder Minima in the future.

—E. N. Parker, University of Chicago (2003)[1]

The many, many thousands of pages of the Assessment Reports of the UN's climate panel, the Intergovernmental Panel on Climate Change (IPCC), are the expression of the beliefs of a small circle of scientists and interested lobbyists who, against all evidence, have convinced themselves that humans are having a dramatic effect on the Earth's climate.[2] The IPCC itself describes its role as 'to assess ... the scientific basis of risk of human-induced climate change, its potential impacts and options for adaptation and mitigation.'[3] In short, before they began to compile their reports, the IPCC took 'human-induced climate change' as their belief, perhaps adding the term 'risk of' as a weak attempt to suggest impartiality.

The 'principles' under which the IPCC operates include the following: 'In taking decisions, and approving, adopting and accepting reports, the Panel, its Working Groups and any Task Forces shall use all best endeavours to reach consensus.'[4] The IPCC's objective of consensus is plainly anti-scientific.

In eleventh century Iraq, Alhazen, justly celebrated in the *ummah wahida* of Islam as one of the founders of the scientific method, wrote that the seeker after truth does not place his faith in any mere consensus, however widespread or venerable. Instead, using his hard-won scientific knowledge, he takes care to verify what he has learned of it. 'The road to the truth,' said Alhazen, 'is long and hard: but that is the road we must follow.'

In my field, the physics of the sun, the IPCC asserts against all evidence that the sun has little influence on climate change. This represents neither a consensus nor an authoritative review of the subject. My own summary of the latest science and evidence on the sun's influence on the climate comes to quite opposite conclusions.[5]

Of the 38 co-authors and three review editors of the IPCC's solar sub-chapter (chapter 8 by Myhre *et al.* 2013),[6] only one is an expert on solar physics. Perhaps not surprisingly, then, the subchapter is shot through with critical errors and serious misrepresentations.

These include:

- Misleading discussion of the sun's radiative forcing. The IPCC authors' formulation of the sun and climate relation in terms of the idealised radiative forcing and feedback scenario missed a great opportunity to highlight the *primary* importance of the non-linear dynamics for the evolution of the earth's orbit around the sun which produces unique and non-repeatable changes in the seasonal distribution of incoming sunlight.
- Concealment of problems in determining absolute total solar irradiance. The IPCC authors failed to alert readers to the fact that great uncertainty in measuring the level and variation in total solar irradiation still exist, to the extent that estimates from several measurements differ by as much as 5 to 10 watts per square metre (Wm^{-2}). The uncertainty is so great that we cannot have confidence in any climatic signals from rising atmospheric CO_2.
- Cherry-picking of the total solar irradiance dataset.[7] The IPCC authors largely ignored at least two other datasets that can be shown to be of better quality.
- Outdated and biased selection of references. The IPCC authors failed, for example, to cite potentially important sun-climate connection paper by Willie Soon and David Legates (2013) which provides disconfirming evidence against the role of atmospheric CO_2 for recent climate change.[8]
- Insufficient understanding of the problems involved in reconstructing total solar irradiance by the method described in Steinhilber et al. in 2009 and 2012.[9] Steinhilber et al.'s method of reconstruction is of very poor, if not incorrect, quality but it is being promoted by IPCC authors to be the best results for historical values of total solar irradiance.
- Misplaced reliance on synthetic eleven-year solar cycles. There are no known measurements to suggest the existence of the eleven-year-like solar cycles in the sun irradiance variation for all historical time. Because the IPCC authors proposed that all paleoclimate modeling group to assume such a cycle in their climate models, their conclusion can create artificial results and misleading conclusions on decadal variations in the actual climate system.

- Ignorance of the 2011 paper by Fontenla et al., which is the best paper on physical modeling of sun's irradiance at all wavelengths.[10] The IPCC authors neither cite Fontenla et al. (2011), nor do they include the important conclusion from Fontenla et al. (2011) that accurate knowledge and information on solar UV irradiance is important for a physical modeling of the relationship between the stratosphere and troposphere.

- Misrepresentation of solar magnetic field measurements by William Livingston and Matt Penn at the U.S. National Solar Observatory. Livingston and Penn clearly suggest that their solar magnetic field observation may be extrapolated to yield a highly weakened sun in the near future, but IPCC authors misrepresented the fact by suggesting only a minor effect.

- Erroneous and unqualified rejection of the study of sun-like stars. The IPCC authors incorrectly claimed that the study of sun-like stars is not important for any physical facts about our sun. Their rejection is based on the illogical assumption that useful knowledge about the sun cannot be obtained from observation of other stars.

The IPCC draws attention to the estimated inter-annual variability of the sun's radiative forcing as being merely several twentieths of a Watt per square meter, but overlooks the 90 Wm^{-2} change that occurs in total solar irradiance from its maximum to its minimum distance from Earth. The relative importance of seasonal insolation on historical climate change can be readily demonstrated. During the warm Eemian interglacial period 130,000 to 110,000 years ago, the amplitude of the seasonal insolation was two to three times larger than today because the eccentricity of the earth's orbit was 4 per cent greater. Even if one wished to assume no difference in intrinsic, non-orbital component of solar irradiance between the Eemian warm period and the current or Holocene warm period, the importance of the relatively large seasonal insolation during the Eemian period in explaining the drastically warmer climate that then prevailed is consistent with many high quality paleoclimate records.

What cannot be ignored is the persistent and systematic failure of computer climate model simulations cited by the IPCC to represent and simulate the full dynamics of the seasonal evolution of climate. In a 2014 paper, Timothy Cronin recently highlighted the fact that even the simple question of what average solar zenith angle to use in climate models is not resolved and that the incorrect representation of solar zenith angle can lead to a surplus of solar radiation of 7 to 20 Wm^{-2} in the global energy budget.[11]

To take an example, the January (i.e. winter minimum) temperatures in China during the mid-Holocene Climate Optimum period of roughly 6000 years ago were known to be about 6 to 8°C warmer than present, based on the study of paleo-vegetation and pollens. In a 2012 paper, Dabang Jiang et al. recently showed that all 36 of the world's best climate models in the Paleoclimate Modeling Intercomparison Project backward forecasted *cooler* winter temperatures for the mid-Holocene than present.[12] In addition, for the mid-Holocene annual-mean temperatures in China, 35 out of 36 models incorrectly simulated a cooler climate than present.[13]

IPCC authors, and the computer models on which they rely, have also arbitrarily and incorrectly preferred the Physikalisch-Meteorologisches Observatorium Davos (PMOD) measurement of total solar irradiance (as exemplified in Claus Fröhlich's 2009 paper) to two independent and arguably better results deduced by the Royal Meteorological Institute of Belgium (RMIB) (Sabri Mekaoui and Steven deWitte, 2008) and the Active Cavity Radiometer Irradiance Monitor (ACRIM) (Nicola Scafetta and Richard Willson, 2014) groups.[14] A 2013 paper by Ali BenMoussa et al. recently raised the important question of accuracy in the calibration of satellite solar instruments, most of which are strongly affected by both in-orbit light and charged-particle radiation exposure and orbital decay.[15] These authors concluded that the better quality control of the RMIB data were preferable to the rather subjective adjustments in the PMOD data—adjustments that were personally specified by Fröhlich, the Principal Investigator of the PMOD instrument. Scafetta and Willson in their 2014 paper provided detailed accounts on the PMOD data adjustment procedure and independently confirm the lower quality of the PMOD data.[16]

Furthermore, IPCC authors have also failed to disclose or to explain that the measurement of total solar irradiance is confounded by our current inability to determine its absolute value.[17] It is surely important to know whether the mean value is 1360, 1361 or 1365 Wm^{-2} because without knowing how the mean climatic state is determined it would be impossible to confirm how the actual climate system is actually changing. The scientific importance of this indeterminacy[18] is also clear if one considers that, according to the IPCC's 2013 *Fifth Assessment Report*, the entire influence of humans on the climate since 1750 is a mere 2.3 Wm^{-2}.[19]

The 2013 paper by Soon and Legates, published before the IPCC's paper cut-off deadline, shows that a reconstructed history of solar irradiance can explain the changes in the Equator-to-Arctic surface temperature gradient from 1850-2012.[20] Scientifically, this result is important for understanding climate dynamics because the Equator-to-Arctic temperature gradient has long been suspected as a key driver of the earth's climate.[21] The IPCC, which purports to review all relevant scientific literature, makes no mention of this important result.

Likewise, it is disturbing that the IPCC has promoted the reconstructed history of total solar irradiance by Steinhilber (2009, 2012) as the best possible result.[22] The Steinhilber reconstruction hinges on very weak statistical links (or lack thereof) between the radial magnetic field variable, B_r, and total solar irradiance published by Fröhlich, a co-author of Steinhilber in 2009.[23] Strictly speaking, the relationship between these two variables shown in Fröhlich's 2009 paper was based *on at most only three data points*, and this one stunning fact should be enough to shy any objective scientist away from attributing importance to the result.[24] Yet the IPCC asserts that changes in the sun's irradiance over the past 1000 to 10,000 years must be small on the basis of Steinhilber's ill-evidenced but convenient result.[25]

The IPCC cites the 2011 paper by Gavin Schmidt et al. as the best and most representative of the state of knowledge on reconstructing total solar irradiance, while condemning as flawed other recent studies of sun-like stars.[26] Schmidt et al. propose that climate models should adopt their novel 1000-year solar irradiance reconstruction which features an artificially imposed eleven-year cycle in total solar irradiance. The current state of knowledge does not support that approach; insufficient evidence exists either in direct measurements or in any reconstruction of total solar irradiance for the past climate.

Selective citation from the scientific literature by the IPCC is clearly evident and its impact is serious. The sun's irradiance covers many wavelengths, primarily from infrared to visible light to ultraviolet. Solar ultraviolet radiation is known to be a very important influence on the amount of ozone in the stratosphere and it controls how much energy large scale planetary waves can carry from the surface and from the climatically-active region of the atmosphere to achieve dynamic equilibrium. It is puzzling that the IPCC's *Fifth Assessment Report* never mentioned the 2011 paper by Fontenla et al., which is the best available scientific paper on the physical basis of, as opposed to the statistical or numerical modeling of, solar irradiance.[27] Fontenla et al. account for many of the magnetic field structures and other features that are observed on the sun and contribute to solar variability.

Furthermore, the IPCC has gravely misrepresented the work of two scientists from the U.S. National Solar Observatory. In papers from 2009 and 2011, Livingston and Penn concluded, on the basis of direct measurement of the trend in the magnetic fields of sunspots, that if current trends were to continue, large sunspots might soon altogether disappear from the face of the sun.[28] Instead, the IPCC in 2013 wrote that Livingston and Penn had suggested that only 'half' of the sunspots might disappear. Specifically, the two solar physicists had written in their 2009 paper, 'A simple linear extrapolation of those data suggested that sunspots might completely vanish by 2015.'[29] Penn and Livingston further commented that 'the predicted dearth in sunspot numbers, independent of the eleven-year sunspot cycle, has proven accurate ... [t]he vigor of sunspots, in terms of magnetic strength and area, has indeed greatly diminished.'[30] The scientific results and conclusion on the current and ongoing magnetic state of our sun by Livingston and Penn are important simply because the variation of the sun's energy output is known to be important driver of Earth climate and that the flat-trending nature of the global temperature for the past fifteen to seventeen years has long been speculated to be connected to the recently weakened sun's magnetic behavior.

Finally, the IPCC asserts that studies comparing sun-like stars to the sun are flawed because the sun has been proven to exhibit atypical variations in magnetic field and brightness. Yet regarding an implication that the sun exhibits a 'normal' level of magnetic activity as compared to the *Kepler* sample of sun-like stars, the latest paper from the NASA's *Kepler* mission[31] by Gibor Basri, Lucianne Walkowicz and Ansgar Reiners asserts:

> We find no empirical evidence in the *Kepler* data for an excess of young active solar-type stars near us, nor is the sun unusually photometrically quiet compared to its neighbors. That is perhaps not surprising, given similar results for Ca II (T. Henry *et al.* 1996) ... There have been previous suggestions that the sun might be photometrically quieter than the bulk of similar stars ... although they were tentative.[32]

The paper by Basri et al. in fact shows that the evidence from the study of sun-like stars suggests a much larger amplitude of solar light output variations than that which has been estimated for our own sun from various satellite projects to date.

In conclusion, the IPCC has been practicing 'para-science' in that, while it affects the appearance of practicing science, it has violated long-held scientific norms and practices of fully and accurately representing the current state of scientific knowledge, and of proposing and testing alternative hypotheses in order to extend knowledge.[33]
Instead, the IPCC and its authors have acted out of prejudice in a manner that has misled both politicians and a largely unsuspecting public.

As a redress, I have spelled out here several of the IPCC's numerous, specific and grievous errors in science. Each error has the effect of minimising the role of the sun and thereby supporting the IPCC's unsupportable claim to be '95 per cent confident' that most of the 0.7°C global warming since 1950 was manmade. That assertion is made without evidence. The assertion is also self-serving, in that the IPCC depends on it for its own continued existence.

Contrary to reports of a '97 per cent consensus', the 2014 paper by Legates et al. demonstrated that only 0.5 per cent of the abstracts of 11,944 scientific papers on climate-related topics published over the 21 years from 1991-2011 had explicitly stated an opinion that more than half of the global warming since 1950 had been caused by human emissions of CO_2 and other greenhouse gases.[34] The overwhelming majority of scientists in climate and related fields, therefore, remain commendably open to the possibility that some other influence—such as the sun—may be the true *primum mobile* of the Earth's climate.

In manipulating its selection and representation of the scientific literature on the solar influence on global mean surface temperature, the IPCC has attempted to bolster the stance of a tiny minority of scientists and then to pretend, with 95 per cent confidence, that this represents a 'scientific consensus'. Such a consensus, even if it did exist, would be of no interest to science.

The central lesson to be learned from this episode in scientific history is that to create an organisation financially and ideologically dependent upon coming to a single, aprioristic viewpoint, regardless of the objective truth, is to create a monster that ignores the truth. Regrettably, the cumulative effect of the IPCC's conduct over the last 25 years has inflicted severe and long-term damage on the reputation of science and of scientists everywhere.

The sun is the ultimate factor in causing change of terrestrial climate. At a small but measurable level, the sun varies, just as most stars do. Centuries of observation and more recent research strongly suggest that our climate is modulated in important ways by the sun's variability. The basic physics of this connection is still poorly understood and stands at the frontier of research. But the body of needed raw data is now available, unlike even twenty years ago. Studying the response and interrelation of other planets to the sun's variability will be extremely helpful in understanding our climate.

5   The scientific context
Robert M. Carter

The science of global warming is at the same time relatively simple and very complex.[1] The simplicity exists in the elemental science that is taught in introductory earth science and meteorology courses at universities around the globe. The complexity lies in the integration of the many and diverse processes involved in climate change, some of which are poorly understood or even remain unknown, thus necessitating the use of speculative computer models to attempt to achieve further understanding.

Those scientists who support the alarmist agenda of human-caused global warming tend to stress the intricacies of climate change, an approach that implies the need for 'experts' and highly complex computer models to adjudicate on the matter. In contrast, independent scientists tend to stress the importance of the broader facts that provide the context against which the threat of a dangerous human influence on climate should be judged.

In this chapter, four basic scientific facts are described that provide an essential context for intelligent discussion of the global warming issue. Thereafter, there is an explanation of the reasons for the differing advice to governments that is provided by the Intergovernmental Panel on Climate Change (IPCC) and the Nongovernmental International Panel on Climate Change (NIPCC). Fuller discussion of these and related issues is provided in my 2010 book, *Climate: the Counter Consensus*.[2]

The context of contemporary climate change

Context 1—error bounds on reconstructing the global average temperature from thermometer data

The main record used by the IPCC for analysing contemporary 'climate change' is compiled by averaging individual temperature records of varying quality and length from around the globe. This, the 'HadCRUT' record—the name being a combination of the data sets of two organisations namely the Hadley Centre of the UK Met Office and the Climatic Research Unit (CRU) at the University of East Anglia—has a number of known deficiencies. These include that it is far too short to be treated as a serious climate record (being equivalent to just five *climate* data points), is probably inadequately corrected for the urban heat island effect, and is subject to other large errors.

For example, amongst the papers released in the 2009 *Climategate* scandal was a previously unpublished 2005 CRU contractual report by Philip Brohan et al. which contained a careful analysis of the likely error bounds for the HadCRUT3 record.[3] These authors considered January 1969 temperature data for measurement and sampling error, temperature bias effects, and the effect of limited observational coverage on large-scale averages. The analysis revealed worldwide errors in the range of 1-5°C for individual sampled area-boxes, i.e. errors that far exceed the total claimed twentieth century warming of ~0.7°C. Clearly, errors for records collected earlier in the twentieth century are likely to be higher still than the already large 1969 errors.

Despite the claim otherwise by Brohan et al. in their 2006 paper, 'Uncertainty estimates in regional and global observed temperature changes', these results indicate that no statistically significant modern warming will be able to be inferred on the basis of HadCRUT or similar thermometer-based records until the current temperature rises over 1°C above that computed for 1969.[4]

Though global average temperature may have warmed during the twentieth century, no direct instrumental records exist that demonstrate any such warming within an acceptable degree of probability.

Context 2—natural temperature variations over geological time

It is a scientific truism that climate persistently changes through time, one manifestation of which is the changing global average temperature recorded in many geological data sets. These data sets are collected, for example, from high latitude ice cores or oceanic seabed cores. Though in the first instance they yield local or regional temperature data, the strong commonality that exists between different palaeoclimate records from widely different geographical regions nonetheless often reflects an underlying global signal.

Because short thermometer temperature records such as HadCRUT manifestly do not comprise an adequate climate record, it is to these geological datasets that we must turn to provide the proper climatic context against which to assess modern temperature changes.

A case in point is the high-quality inferred air temperature above the Greenland ice cap for the last 10,000 years (Figure 1). This record shows (i) that temperatures were up to a full two degrees warmer than today during the Holocene Climatic Optimum, c. 8,000 years before present (BP); (ii) the presence of a persistent millennial cycle of warmings and coolings, with all pre-modern peaks of this cycle, including the Mediaeval Warm Period, being warmer than the late twentieth century peak;[5] and (iii) an overall cooling of temperature since 8,000 years BP which took place against the background of an increase of atmospheric carbon dioxide (CO_2) of 20 ppm (of natural origin, and as recorded in Antarctic ice cores).

Figure 1: Greenland air temperature: last 10,000 years

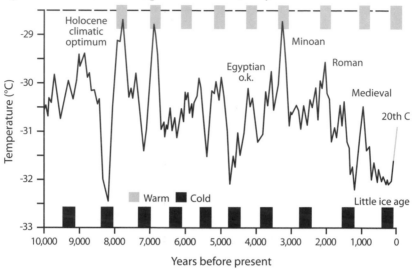

Greenland surface air-temperatures over the Holocene (last 10,000 years) as revealed by proxy measurements of oxygen isotope ratios. Note that all peaks prior to the Little Ice Age were warmer than the late twentieth century warming.

Source: R. B. Alley, "The Younger Dryas cold interval as viewed from central Greenland," *Quaternary Science Reviews*, Vol. 19 (2000): 213-226.

Though the Greenland average temperature has warmed by about 1°C since the Little Ice Age, such warming cannot be shown to have resulted from increases in human-related CO_2 emissions. The warming was also entirely unalarming in rate and magnitude when compared with other similar natural warmings that occurred over the preceding 10,000 years.

Context 3—carbon dioxide variations over geological time

It is widely misrepresented in the public domain that Earth's current levels of atmospheric CO_2 are dangerously and atypically high. Such claims are false, because modern CO_2 levels lie near to an all-time low as assessed against the geological record.

500 million years ago, before land-plant photosynthesis was operating, atmospheric CO_2 attained about fifteen times present day levels, built by cumulative volcanic exhalations during four billion years of pre-Cambrian time. Between about 450 and 320 million years ago, levels declined steadily from >4,000 parts per million (ppm) to around 500 ppm, concomitant with the evolution and diversification of land plants, and thereafter varied between about 500 and 2000 ppm. In effect, since about 450 million years ago plant photosynthesis has removed CO_2 from the atmosphere, after which normal sedimentary burial processes led to the conversion of original vegetation to coal and thus the storage of former CO_2 underground. Utilising coal as an energy resource simply returns the CO_2 to the atmosphere from whence it came in the first place, yielding the twin benefits of generation of cheap electricity and the greening of the planet. Why radical environmentalists view this as a problem has never been explained adequately, and remains mysterious to this day. For at 280 ppm (pre-industrial), 400 ppm (today), 560 ppm (2 x pre-industrial) or even 1120 ppm (4 x pre-industrial), planet Earth's atmosphere would still remain in a CO_2-starved state.

Carbon dioxide is an essential trace-gas for plant photosynthesis, and therefore vital for biodiversity and the sustenance of most planetary food chains. Adding carbon dioxide to the atmosphere is beneficial for the growth of many plants, especially cereals, and by comparison with the last 550 million years of geological history, Earth currently exists in a state of carbon dioxide starvation.[6]

Context 4—efficacy of warming caused by extra carbon dioxide

Carbon dioxide is a potent greenhouse gas for intercepting space-bound (and hence cooling) radiation emitted from Earth's surface at wavelengths around 14.8μ and 9μ within the electromagnetic spectrum.

Initially, at low atmospheric concentrations, the gas therefore has a strong greenhouse effect as it blocks outgoing radiation at these wavelengths. However, the narrowness of the spectral intervals across which CO_2 intercepts radiation results in a rapid saturation of its effect, such that every doubling in the concentration of CO_2 enhances the greenhouse effect by a constant amount. This is reflected as the negative logarithmic relationship that exists between extra CO_2 and the warming that it causes.

Because of this logarithmic relationship, the amount of warming caused by increasing quanta of CO_2 depends upon the level of CO_2 already in the atmosphere, and diminishes steadily in a 'less-temperature-bang-for-every-incremental-carbon-dioxide-buck' pattern. Given the pre-industrial starting point of 280 ppm of atmospheric CO_2, only minor additional warming will occur in response to the much-feared doubling of CO_2 to 560 ppm.

While scientists generally agree that this *prima facie* warming will cumulate to about 1°C for a doubling, IPCC scientists allege that the positive feedback effect from more water vapour, itself generated by the initial CO_2-forced warming, will result in a total warming of about 3-6°C. However, this speculation conflicts with other empirical data and is therefore controversial.

Though CO_2 is a greenhouse gas, its warming efficacy rapidly diminishes (in logarithmic fashion) as atmospheric concentrations rise.

When both positive (e.g. enhanced water vapour) and negative (e.g. enhanced low level cloud) feedback effects, and geological climate records, are taken into consideration, little likelihood exists that conceivable levels of human emissions will cause dangerous future warming.

Common ground amongst global warming protagonists

These four contextual scientific matters discussed earlier all point to a lack of alarm regarding dangerous global warming caused by human-related greenhouse gas emissions, despite which a vigorous public debate about the matter continues. This debate is strongly antagonistic, belying the fact that a large measure of scientific agreement exists between the debating protagonists about most of the facts of the matter.

The common scientific ground and the main matters that remain in dispute are summarised in Table 1.

Table 1: The essence of the scientific debate

The common ground includes:
- That climate has always changed and always will.
- That CO_2 is a greenhouse gas, the accumulation of which results in warming of the lower atmosphere.
- That post-industrial human-related emissions comprise a new source of atmospheric CO_2
- That a global warming of around 0.7°C may have occurred in the twentieth century.
- That global warming has ceased over the last seventeen years.

The three key science issues remaining in dispute are:
- The amount of net warming that is, or will be, produced by the 'extra' human-related emissions.*
- Whether any actual evidence exists for dangerous warming of human causation over the last 50 years.
- Whether the IPCC's computer models can provide accurate climate predictions 100 years into the future.

Usually termed the 'climate sensitivity' issue, which equates in turn with the amount of warming that will be produced by a doubling of carbon dioxide over its pre-industrial level.

Why is this common ground not more generally understood? The answer is simple. Since the formation of the IPCC in 1988, a wide-ranging and worldwide propaganda campaign has been conducted to raise public alarm about global warming.[7] Though initially promulgated by environmental organisations, commercial lobbyists (e.g. the wind power industry), and the financial markets, the global warming bandwagon soon attracted the attention of politicians because of the electoral advancement that it promised, and has all the while been egged-on by a ceaselessly alarmist press corps.

All the classic tools of propaganda and spin have been deployed for the advancement of public alarm about global warming, including scientific malfeasance, noble cause corruption, the makeover of formerly independent expert groups such as academies of science, the indoctrination of school children from kindergarten onwards and the *ad hominem* demonisation of scientists who fail to conform to the orthodox IPCC view. The many hundreds of risible claimed negative effects of global warming (many of which are actually beyond parody) are summarised in a hilarious list of claimed warming misadventures compiled by UK engineering professor, John Brignell.[8]

During the 1990s and the first few years of the present century, public opinion was visibly affected by this barrage of pseudo-scientific propaganda, as reflected by a clear majority of the citizens of OECD countries regularly expressing their concern in opinion polls. Over the last ten years, however, and thanks not a little to the assiduous efforts of independent scientists and organisations such as the NIPCC, public opinion has swung away from the global warming scare and other similarly over-hyped environmental causes. At the same time, many cartoonists and comedians have started to lampoon the more ridiculous claims of the global warming alarmists. As one expressed it:

> So just when those supporting climate alarm thought that they had everything settled and nailed down, a gale of discontent started to blow. Cartoonist heaven really. We love the spectacle of powerful people preparing their policy against strong winds and rough seas, frantically rigging up fragile, flapping sails of spin and blather. If you're going to spend over $15 billion of taxpayer's money on desalinated water, or manage a potentially

ruinous carbon dioxide trading scheme… then you certainly don't want to be questioned too closely, let alone lampooned, about the scientific details that you misunderstood or got wrong.[9]

Science should not be about emotion or politics, yet it is uncomfortably true that public discussion of the global warming issue has for many years been conducted far more in accordance with those criteria than it has been concerned with science per se. There are three prime reasons for this.

First, as a branch of the United Nations, the IPCC is itself an intensely political and not a scientific body. As its chairman, Dr Rajendra Pachauri observed in an interview with the *Guardian* newspaper:

> We are an intergovernmental body and we do what the governments of the world want us to do. If the governments decide we should do things differently and come up with a vastly different set of products we would be at their beck and call.[10]

To boot, the IPCC charter requires that the organisation investigates not climate change in the round, but solely global warming caused by human greenhouse emissions, a blinkered approach that consistently damages all IPCC pronouncements.

Second, from local green activist groups up to behemoth NGOs like Greenpeace and WWF, over the last twenty years the environmental movement has espoused saving the planet from global warming as its *leitmotif*. This has had two devastating results. One is that radical environmentalists have worked relentlessly to sow misinformation about global warming in both the public domain and the education system. And the other is that, faced with this widespread propagandisation of public opinion and young persons—and also by strong lobbying from powerful self-interested groups like government research scientists, alternative energy providers and financial marketeers—politicians have had no choice but to fall into line. Whatever their primary political philosophy, all active politicians are daily mindful of the need to assuage the green intimidation and bullying to which they and their country's industries and citizens are incessantly subjected—and to which no Western country has yet devised a feasible counter.

Third, and perhaps most influential of all, with very few exceptions major media outlets have provided unceasing support for measures to 'stop global warming'. This behaviour appears to be driven by a combination of the left-wing ('liberal' in the US sense) and green personal beliefs of most reporters, and the commercial nose of experienced editors who understand that alarmist environmental reporting sells both product and advertising space. As one experienced editor has written:

> The publication of 'bad news' is not a journalistic vice. It's a clear instruction from the market. It's what consumers, on average, demand.... As a newspaper editor I knew, as most editors know, that if you print a lot of good news, people stop buying your paper. Conversely, if you publish the correct mix of doom, gloom and disaster, your circulation swells. I have done the experiment.[11]

Where to from here—the IPCC and NIPCC Reports

And thus we arrive at the present impasse, in which the IPCC and its attached covey of special interest groups continue to argue vehemently in favour of taking costly action to limit industrial CO_2 emissions at the same time that many thousands of qualified independent scientists assert that government policy should be concerned with adaptation to natural climate hazard rather than targeting chimerical human-caused warming. Faced with this conflicting advice, some western governments are continuing to respond to IPCC alarmism by taking penal financial measures against CO_2 emissions (e.g. USA, UK) whereas others have already signalled firmly that they are not prepared to enter into new Kyoto-style anti-emissions agreements (e.g. Canada, Japan, Australia).

To attain a fuller and mature understanding of the topic, and to move the matter forward, requires that policymakers consult and compare the two compendious and up-to-date summaries of climate-related research that are produced respectively by the IPCC[12] and the NIPCC.[13] Both teams of authors provide similar scholarly analysis and summary of recent scientific papers, but with the key difference that whereas IPCC scientists are effectively government-appointed and work in close cooperation with environmental lobbying agencies,[14] NIPCC scientists are fully independent of political, financial and lobby group influences.

Given this difference, and despite the fact that the corpus of scientific papers they consider is similar, it is perhaps unsurprising that the scientists of the IPCC and NIPCC have reached diametrically opposing conclusions about the hazard posed by human-related CO_2 emissions, as summarised in accompanying Tables 2 and 3.

Table 2: Selected primary science conclusions—IPCC 2013

- Warming of the climate system is unequivocal, and since the 1950s, many of the observed changes are unprecedented over decades to millennia. The atmosphere and ocean have warmed, the amounts of snow and ice have diminished, sea level has risen, and the concentrations of greenhouse gases have increased.

- Ocean warming dominates the increase in energy stored in the climate system, accounting for more than 90 per cent of the energy accumulated between 1971 and 2010. It is virtually certain* that the upper ocean (0-700 m) warmed from 1971 to 2010, and it likely warmed between the 1870s and 1971.

- Over the last two decades, the Greenland and Antarctic ice sheets have been losing mass, glaciers have continued to shrink almost worldwide, and Arctic sea ice and Northern Hemisphere spring snow cover have continued to decrease in extent.

- The rate of sea level rise since the mid-nineteenth century has been larger than the mean rate during the previous two millennia. Over the period 1901–2010, global mean sea level rose by 0.19 [0.17 to 0.21]m.

- The atmospheric concentrations of CO_2, methane, and nitrous oxide have increased to levels unprecedented in at least the last 800,000 years. CO_2 concentrations have increased by 40 per cent since pre-industrial times, primarily from fossil fuel emissions and secondarily from net land use change emissions. The ocean has absorbed about 30 per cent of the emitted anthropogenic CO_2, causing ocean acidification.

- Human influence on the climate system is clear. This is evident from the increasing greenhouse gas concentrations in the atmosphere, positive radiative forcing, observed warming, and understanding of the climate system.

- Human influence has been detected in warming of the atmosphere and the ocean, in changes in the global water cycle, in reductions in snow and ice, in global mean sea level rise, and in changes in some climate extremes. It is extremely likely that human influence has been the dominant cause of the observed warming since the mid-twentieth century.

- Cumulative emissions of CO_2 largely determine global mean surface warming by the late twenty-first century and beyond. Most aspects of climate change will persist for many centuries even if emissions of CO_2 are stopped. This represents a substantial multi-century climate change commitment created by past, present and future emissions of CO_2.

Probability terms such as "likely", "extremely likely" are italicized in the original IPCC report, to imply statistical significance. Because the phrases are actually devoid of statistical or scientific meaning such italics have not been reproduced in this summary table.

Table 3: Primary science conclusions—NIPCC 2013, 2014

Physical science

- Neither the rate nor the magnitude of the reported late twentieth century surface warming (1979–2000) lay outside normal natural variability, nor was it in any way unusual compared to earlier episodes in Earth's climatic history. Furthermore, solar forcings of temperature change are likely more important than is currently recognised, and evidence is lacking that a 2°C increase in temperature (of whatever cause) would be globally harmful.

- No unambiguous evidence exists for adverse changes to the global environment caused by human-related CO_2 emissions. In particular, the cryosphere is not melting at an enhanced rate; sea-level rise is not accelerating; no systematic changes have been documented in evaporation or rainfall or in the magnitude or intensity of extreme meteorological events; and an increased release of methane into the atmosphere from permafrost or sub-seabed gas hydrates is unlikely.

- The current generation of general circulation climate models (GCMs) are unable to make accurate projections of climate even ten years ahead, let alone the 100-year period that has been adopted by policy planners. The output of such models should therefore not be used to guide public policy formulation until they have been validated and shown to have predictive value.

Biological impacts

- Atmospheric CO_2 is not a pollutant. It is a non-toxic, non-irritating, and natural component of the atmosphere. Long-term CO_2 enrichment studies confirm the findings of shorter-term experiments, demonstrating numerous growth-enhancing, water-conserving, and stress-alleviating effects of elevated atmospheric CO_2 on plants growing in both terrestrial and aquatic ecosystems.

- The ongoing rise in the air's CO_2 content is causing a great greening of the Earth. All across the planet, the historical increase in the atmosphere's CO_2 concentration has stimulated vegetative productivity. This observed stimulation, has occurred in spite of many real and imagined assaults on Earth's vegetation, including fires, disease, pest outbreaks, deforestation, and climatic change.

- There is little or no risk of increasing food insecurity due to global warming or rising levels. Farmers and others who depend on rural livelihoods for income are benefitting from rising agricultural productivity throughout the world, including in parts of Asia and Africa where the need for increased food supplies is most critical. Rising temperatures and atmospheric CO_2 levels play a key role in the realisation of such benefits.

- Terrestrial ecosystems have thrived throughout the world as a result of warming temperatures and rising levels of atmospheric CO_2. Empirical data pertaining to numerous animal species, including amphibians, birds, butterflies, other insects, reptiles, and mammals, indicate global warming and its myriad ecological effects tend to foster the expansion and proliferation of animal habitats, ranges, and populations, or otherwise have no observable impacts one way or the other. Multiple lines of evidence indicate animal species are adapting, and in some cases evolving, to cope with climate change of the modern era.

- Rising temperatures and atmospheric CO_2 levels do not pose a significant threat to aquatic life. Many aquatic species have shown considerable tolerance to temperatures and CO_2 values predicted for the next few centuries, and many have demonstrated a likelihood of positive responses in empirical studies. Any projected adverse impacts of rising temperatures or declining seawater and freshwater pH levels ('acidification') will be largely mitigated through phenotypic adaptation or evolution during the many decades to centuries it is expected to take for pH levels to fall.

- A modest warming of the planet will result in a net reduction of human mortality from temperature-related events. More lives are saved by global warming via the amelioration of cold-related deaths than those lost under excessive heat. Global warming will have a negligible influence on human morbidity and the spread of infectious diseases, phenomena observed in virtually all parts of the world.

Thoughtful analysis of the IPCC and NIPCC summaries of the scientific literature (Tables 2, 3) reveals (i) a lack of empirical evidence for human-caused global warming, (ii) that the temperature fluctuations that occurred in the twentieth century fell well within previous natural bounds, and reinforces the fact (iii) that IPCC's advice about future dangerous warming is entirely predicated upon the accuracy of their speculative computer models. In addition to which, the four contextual tests outlined in the earlier part of this essay also provide strong evidence against alarm.

Some will say, nonetheless, that given that the science lacks certainty (whatever that might mean) we should give Earth the 'benefit of the doubt', by which they mean taking precautionary action against human-related CO_2 emissions, just in case they should cause dangerous warming. However, this catchy phrase reveals a profound misunderstanding of the real climatic risks faced by our societies, because it assumes that global warming is more dangerous, or more to be feared, than is the equally likely occurrence of global cooling; in reality, the converse is more likely to be true.[15]

It needs to be recognised that the theoretical hazard of dangerous human-caused warming is but one small part of a much wider climate hazard that all scientists agree upon, which is the dangerous weather and climatic events that Nature intermittently presents us with—and always will. It is clear from the many and continuing climate-related disasters that occur around the world that the governments of even advanced, wealthy countries are often inadequately prepared for such disasters. We need to do better, and squandering money to give Earth the benefit of the doubt based upon an unjustifiable assumption that dangerous warming will shortly resume is exactly the wrong type of 'picking winners' approach.

Many scientists, including leading solar physicists, currently argue that solar cycling implies that the most likely climatic trend over the next several decades is one of significant cooling rather than warming. Meanwhile, the IPCC's computer modellers assure us with all the authority at their command that global warming will shortly resume—just you wait and see.

The reality is that no scientist on the planet can tell you with credible probability whether the climate in 2030 will be cooler or warmer than today. In such circumstances the only rational conclusion to draw is that we need to be prepared to react to either warming or cooling over the next several decades, and also to severe weather events, depending upon what Nature chooses to serve up to us. A primary government duty of care is to protect the citizenry and the environment from the ravages of natural climate-related events. What is needed is not unnecessary and penal measures against CO_2 emissions, but instead a prudent and cost-effective policy of preparation for, and adaptive response to, all climatic events and hazards.

The appropriate response to climate hazard, then, is for national policies to be based on preparing for and adapting to all climate-related events as and when they happen, and irrespective of their presumed cause. Every country needs to develop its own understanding of, and plans to cope with, the unique combination of climate hazards that apply within its particular boundaries. The planned responses should be based upon adaptation, with later mitigation where appropriate to cushion citizens who are affected in an undesirable way.

The idea that there can be a one-size-fits-all global solution to deal with just one possible aspect of future climate change, as recommended by the IPCC and still favoured by green activists and most media commentators, fails entirely to deal with the real climate and climate-related hazards to which we are all exposed every day.

6 Forecasting rain

John Abbot & Jennifer Marohasy

A reasonable test of the value of any scientific theory is its utility. For example, the calendars that were developed based on Nicolaus Copernicus' Heliocentric Theory of the Universe were better calendars than those based on Ptolemy's Handy Tables. The new calendars, based on the new theoretical approach that suggested the earth moved around the sun (rather than the sun around the earth), more precisely predicted the position of the sun and the planets and thus was of more utility for navigation and also weather forecasting. The adoption of the new theory, however, was resisted because of its political and religious implications. In contrast, the theory of anthropogenic global warming (AGW) has very little practical utility, but tremendous political value. It is a theory that accords with the mood of our time, the zeitgeist, which assumes that man's greed is despoiling the earth and that political action based on a scientific consensus can save the planet.

This is part of the reason why credible scientific rebuttals fail to achieve its overthrow. The other reason is that the history of science shows it is competition, not logical argument, that only ever successfully displaces even a failed paradigm. For example, in a review of seven high-profile scientific articles and also devastating rebuttals some of which were also published in the journals *Science* and *Nature*, Jeannette

Banobi and colleagues showed that original articles were cited seventeen times more than their rebuttals, and that annual citations of the original articles were unaffected by the rebuttals.[1] Indeed it is a naïve view of science that assumes new hypotheses are either accepted or rejected according to the received evidence. As Thomas Kuhn explains in his 1962 seminal work *The Structure of Scientific Revolutions*, it is more usual that an established paradigm will be continually modified in small ways to incorporate contrary evidence—that is, until a new theory emerges with the potential to capture the imagination of the discipline, or at least in the first instance, a critical 10 per cent.[2]

The status quo

Prior to the establishment of the current Australian Bureau of Meteorology in 1909, Australian meteorologists had a keen knowledge of astronomy and considered solar, lunar and planetary cycles in their weather forecasting. There remained some interest in this approach, which was termed solar terrestrial physics, at the bureau until the early 1950s. Since the 1950s the bureau, and other major climate research institutions around the world, have worked towards a global effort to simulate climate largely independently of extraterrestrial influences. The primary focus has instead been modeling oceanic and atmospheric processes on earth as an array of thousands of numbers while averaging spatiotemporal variations from outer space, even variations in solar radiation. By the mid-1970s computing power was catching up with the ambition of these scientists to simulate earth's climate and by the 1980s there was a growing confidence in the models and, in particular, their ability to forecast the impact of increasing levels of atmospheric carbon dioxide (CO_2) on climate.

The focus now is on the interpretation of output from general circulation models (GCMs). This is what concerns mainstream meteorologists, with key drivers of climate change thought to relate to human activity. Indeed, the idea that the moon influences the weather through its gravitational effect is generally scoffed at. While meteorology has moved away from a deep knowledge of astronomy, there is no evidence to suggest that the skill of the discipline at weather and climate forecasting, particularly at medium and long-range rainfall forecasting, has improved. There is considerable evidence to suggest it is in decline, despite ever-increasing investments in supercomputers and media releases, not to mention high profile collaborations through the United Nations.

Professor Chris Turney is a modern climate scientist who set off on an expedition to the Antarctic in 2013 believing the GCMs that have been forecasting melting polar ice for some time. Turney subsequently attracted worldwide attention when on Christmas Eve, the Russian-flagged ship he hired got stuck in sea ice. If, before setting out, he had consulted the long-range weather forecasters who operate independently of the established institutions, and without the aid of GCMs but with reference to patterns and phase changes associated with solar and lunar phenomena, he could have been forewarned of the unusually slow melt rate of Antarctic ice last austral summer. For example, Kevin Long is a long-range weather forecaster based in Bendigo, Victoria. When he issued his seasonal forecast at the end of August 2013, it stated that the higher sea ice averages will become one of the dominant drivers of eastern Australia's developing mega-drought cycles as higher sea ice periods go hand-in-hand with below average rainfall and heavier late season frosts. But who listens to Long and other long range weather forecasters, including Ken Ring in New Zealand and Joseph D'Aleo in the United States, who operate independently of the established institutions?

While Australian taxpayers invested upward of 30 million dollars in just one supercomputer in March 2009 on the basis that this would make weather predictions more accurate, Long, without any government support, is arguably producing more reliable medium-term rainfall forecasts. For example, on 28 August 2013 the Bureau issued its national rainfall outlook for the austral spring of 2013 indicating that it would be wetter than average for most of the Murray Darling Basin and especially central Victoria. Managers of major water infrastructure prepared for possible flooding. Long also issued a forecast on 28 August 2013, but his had a dramatically different outlook. Long forecasted below-average rainfall for central Victoria, above-average temperatures and with comments that river stream flows would drop away quickly. Long's forecast was remarkably accurate. On 25 September the Bureau was forced to issue a revised spring forecast, this time forecasting below average rainfall for the Murray Darling Basin.

The Bureau's forecasts are derived from the Predictive Ocean Atmosphere Model for Australia (POAMA). The output from POAMA is generally consistent with what is known of rainfall forecasts from other GCMs. They fail to reproduce the observed historical annual and seasonal mean rainfalls, across southeast Australia.[3] In fact, they are often worse than the forecasts a school child could generate based on simply calculating the monthly mean rainfall for a particular locality with a pencil and pad. Such an average value is known as climatology. But the scientists at the Bureau are so committed to GCMs that in May 2013 they discarded the old statistical models relied on for the past twenty years, and adopted POAMA as the basis for all climate forecasting. There is no peer-reviewed paper that indicates POAMA can reliably produce a seasonal or monthly forecast with more skill than the old statistical method—because it can't. Indeed the peer-reviewed literature, including a new paper by us,[4] shows that statistical models that rely on pattern analysis will consistently outperform GCMs such as POAMA.

Meanwhile many climate scientists seek to minimise the potential error of their GCMs forecasts by promoting average values from multiple models. These are often referred to as ensembles. The *Fifth Assessment Report* (AR5) of the Intergovernmental Panel on Climate Change (IPCC), in its conclusions on 'impacts, adaptation and vulnerability' released in March 2014, included rainfall forecasts for almost every region of the world. The forecasts were developed by averaging output from more than 50 different GCMs. It's a mammoth undertaking and so it's done in stages with results published in the peer-reviewed literature. If we consider results just for the Murray Darling Basin in terms of output from 27 different models run by key institutions across the globe including the Bureau, each contributing to the fifth phase of the Climate Model Intercomparison Project, it is evident that some models predict a large increase in rainfall, while others predict a large overall decline in rainfall for the Murray Darling.[5] The extent of the divergence in output from the individual GCMs should be reason for great concern, and for the discarding of the models altogether, but instead the scientists choose to combine the irreconcilable and thus arrive at a single figure claiming a two per cent decline in rainfall for the Murray Darling by 2090!

Some claim Ken Ring is running a weather prediction scam because he uses the moon to inform his rainfall forecasts. Yet it is not disputed by those with an understanding of conventional physics that the moon's gravitational field, along with the day and night cycle generated by the spinning earth, creates atmospheric tides that modulate high-altitude winds that have a major influence on weather. We have seen no independent assessment of the skill of Ring's predictions, but he sells many hundreds of his weather almanacs to Australian farmers each year. Indeed if there is a scam, it is being run by the Bureau, not Ring, and at the expense of Australian taxpayers with the approval of both the Australian government and the United Nations and the adulation of the Australian Broadcasting Corporation.

Interestingly, POAMA also receives significant direct industry funding including from the Grains Research and Development Corporation and Queensland Canegrowers Ltd. Meanwhile, members of these organisations buy Ring's almanacs.

A test for the existence of natural climate cycles

In his book *Climate: The Counter Consensus*, Professor Robert M. Carter laments that the IPCC concentrates its analyses of climate change on only the last few hundred years, and has repeatedly failed to give proper weight to the geological context of the relatively short 150-year long instrumental record.[6] Carter correctly draws attention to the natural cycles that span tens of thousands of years including changes affected by Earth's tilt (41,000 year), eccentricity (100,000) and precession (20,000). Carter goes on to mention other cycles of shorter duration including those affected by variations in the intensity of the Sun's magnetic fields in particular the Schwabe (eleven year), Hale (22 year) and Gleissberg (70-90 year) periodicities and the effect of the moon and the sun through the 18.6-year-long lunar nodal cycles which he states causes variations in atmospheric pressure, temperature, rainfall, sea-level and ocean temperature, especially at high latitudes.

Long-range US weather forecaster Joseph D'Aleo uses changes in Pacific Ocean temperatures associated with the El Niño-Southern Oscillation (ENSO), and also the multi-decadal Pacific Decadal Oscillation (PDO). He says ENSO is modulated by the PDO, affecting spring floods and tornadoes in the US. Ring explains the southern oscillation in terms of water 'sloshing' from one side of the Pacific Ocean to the other and back again, depending on changing lunar declination.

While essentially ignoring the likely extra-terrestrial origin of the phenomena, the mainstream climate science community does take a keen interest in sea surface temperature anomalies particularly in the Pacific and in particular ENSO. Recognising the global impact of this southern oscillation, but not the moon as a driver, the government-funded scientists have been attempting to simulate ENSO using GCMs. Despite a concerted effort and thousands of peer-review publications on the subject, the skill of the GCMs at forecasting ENSO remains poor; comparable to what could be achieved using the simple statistical models popular 30 years ago.[7] Arguably, the best forecasts of ENSO come not from GCMs or simple statistical models, but from artificial neural networks (ANNs).[8] ANNs are massive, parallel-distributed, information-processing systems with characteristics resembling the biological neural networks of the human brain. They are a form of artificial intelligence and represent state of the art statistical modelling. In contrast to GCMs that attempt to simulate and understand climate from first principles, ANNs simply mine historical data for patterns. Many leaders within the mainstream climate science community are dismissive of ANNs and their application, claiming that because the climate is on a new trajectory, statistical models, including ANNs, are no longer applicable. But if through mining historical records ANNs can produce better ENSO and also rainfall forecasts, this in itself is evidence that natural climate cycles are still operating and that the climate is not on a radically new trajectory.

Output from ANNs and GCMs can be easily and objectively measured using root mean square error (RMSE). This number simply adds together the difference between observed and forecast sea surface temperatures or rainfalls with the bigger the number the worse the forecast. So it's easy to show in an objective way that ANNs can provide a much better medium term ENSO and rainfall forecast. The difficulty has been in generating interest in this approach and interest in the potential of ANNs to revolutionise climate science.

The future

During the 1970s and 1980s, Western democracies moved away from their traditional methods of funding science. In Australia there was a move to project-based funding of a limited duration. So scientists had to identify problems and promote these problems if they were to secure funding—if they were to keep their jobs. At the same time there was increasing interest and reliance in environmental and climate science on mathematical models facilitating a new preference for virtual over observational data. In his book *Science and Public Policy*, Professor Aynsley Kellow of the University of Tasmania, explains how this has also contributed to the hijacking of science by people grinding axes on behalf of noble causes.[9] Richard Lindzen, Professor of Atmospheric Sciences at the Massachusetts Institute of Technology, has repeatedly explained that climate science has now become a source of authority rather than a mode of inquiry, that it has a global constituency and has successfully co-opted almost all of institutional science.

Our work showing the potential application of artificial intelligence to medium-term rainfall forecasting, recently generating five scientific publications, has been possible as a consequence of philanthropic funding from the B. Macfie Family Foundation. The foundation was established to provide scholars with an opportunity to seek out empirical evidence without ideological or commercial interference. Such criteria are often critical to the realisation of scientific research where the benefits of a line of enquiry are not always immediately obvious and are often unpredictable. This investment needs to be leveraged into something more significant if the big questions about natural climate cycles are to be explored: if our initial investigations are to contribute to the development of a new paradigm.

Confirmation bias is a tendency for people to treat data selectively and favor information that confirms their beliefs. This can lead to illusions of invulnerability and belief in the inherent morality of the group leading to self-censorship, illusions of unanimity and an incomplete consideration of alternative solutions to the issue at hand. All of these characteristics can be applied to both the mainstream climate science community and also global warming sceptics. Indeed both groups are convinced of the inherent moral good in their cause and approach to the issue of global warming. But the community at large might be seen as losing interest in the posturing from both camps. Of perhaps more interest than the magnitude of recent increases in temperatures or not, might be a good medium-term weather forecast.

There are natural climate cycles as described by Carter in his book *Climate: The Counter Consensus*. They cycles have an extraterrestrial origin and provide a physical basis for the long-range weather forecasts sold by Ring and D'Aleo. There is a potential role for ANNs in elucidating these relationships and describing them mathematically. Indeed artificial intelligence has the capacity to underpin a new theory of climate in the same way that GCMs currently underpin AGW. ANNs, if run on supercomputers, could revolutionise our capacity to unravel and understand natural climate cycles. Obviously this has direct application for better weather and climate forecasting with immense utility for humanity.

The reality is that those who would like to see AGW theory discarded should increase their expectations of climate science rather than, as many do, support the myth that weather and climate are essentially chaotic and therefore unpredictable. The real long-term solution is philanthropic funding of alternative research programs, funding that backs diversity and competition. The Enlightenment happened because brave men went in search of knowledge. They sought to understand the natural world, not to save it. What climate science needs right now are new tools to replace the failed GCMs, and a new unifying theory of climate to replace the failed theory of anthropogenic global warming.

The politics
and economics
of climate change

7 Cool it: an essay on climate change

Nigel Lawson

There is something odd about the global warming debate—or the climate change debate, as we are now expected to call it, since global warming has for the time being come to a halt.[1] I have never shied away from controversy, nor—for example, as Chancellor—worried about being unpopular if I believed that what I was saying and doing was in the public interest.

But I have never in my life experienced the extremes of personal hostility, vituperation and vilification which I—along with other dissenters, of course—have received for my views on global warming and global warming policies.

For example, according to the Climate Change Secretary Ed Davey, the global warming dissenters are, without exception, 'wilfully ignorant' and in the view of the Prince of Wales we are 'head-less chickens'. Not that 'dissenter' is a term they use. We are regularly referred to as 'climate change deniers', a phrase deliberately designed to echo 'Holocaust denier'—as if questioning present policies and forecasts of the future is equivalent to casting malign doubt about a historical fact.

The heir to the throne and the minister are senior public figures, who watch their language. The abuse I received after appearing on the BBC's Today programme in February 2014 was far less restrained. Both the BBC and I received an orchestrated barrage of complaints to the effect that it was an outrage that I was allowed to discuss the issue on the programme at all. And even the Science and Technology Committee of the House of Commons shamefully joined the chorus of those who seek to suppress debate.

In fact, despite having written a thoroughly documented book about global warming more than five years ago, which happily became something of a bestseller, and having founded a think tank on the subject—the Global Warming Policy Foundation—the following year, and despite frequently being invited on *Today* to discuss economic issues, this was the first time I had ever been asked to discuss climate change. I strongly suspect it will also be the last time.

The BBC received a well-organised deluge of complaints—some of them, inevitably, from those with a vested interest in renewable energy—accusing me, among other things, of being a geriatric retired politician and not a climate scientist, and so wholly unqualified to discuss the issue.

Perhaps, in passing, I should address the frequent accusation from those who violently object to any challenge to any aspect of the prevailing climate change doctrine, that the Global Warming Policy Foundation's non-disclosure of the names of our donors is proof that we are a thoroughly sinister organisation and a front for the fossil fuel industry.

As I have pointed out on a number of occasions, the Foundation's Board of Trustees decided, from the outset, that it would neither solicit nor accept any money from the energy industry or from anyone with a significant interest in the energy industry. And to those who are not—regrettably—prepared to accept my word, I would point out that among our trustees are a bishop of the Church of England, a former private secretary to the Queen, and a former head of the Civil Service. Anyone who imagines that we are all engaged in a conspiracy to lie is clearly in an advanced stage of paranoia.

The reason we do not reveal the names of our donors, who are private citizens of a philanthropic disposition, is in fact pretty obvious. Were we to do so, they, too, would be likely to be subject to the vilification and abuse I mentioned earlier. And that is something which, understandably, they can do without.

That said, I must admit I am strongly tempted to agree that, since I am not a climate scientist, I should from now on remain silent on the subject—on the clear understanding, of course, that everyone else plays by the same rules. No more statements by Ed Davey, or indeed any other politician, including Ed Milliband, Lord Deben and Al Gore. Nothing more from the Prince of Wales, or from Lord Stern. What bliss!

But of course this is not going to happen. Nor should it; for at bottom this is not a scientific issue. That is to say, the issue is not climate change but climate change alarmism, and the hugely damaging policies that are advocated, and in some cases put in place, in its name. And alarmism is a feature not of the physical world, which is what climate scientists study, but of human behaviour; the province, in other words, of economists, historians, sociologists, psychologists and—dare I say it—politicians.

And *en passant*, the problem for dissenting politicians, and indeed for dissenting climate scientists for that matter, who certainly exist, is that dissent can be career-threatening. The advantage of being geriatric is that my career is behind me: there is nothing left to threaten.

But to return: the climate changes all the time, in different and unpredictable (certainly unpredicted) ways, and indeed often in different ways in different parts of the world. It always has done and no doubt it always will. The issue is whether that is a cause for alarm—and not just moderate alarm. According to the alarmists it is the greatest threat facing humankind today: far worse than any of the manifold evils we see around the globe which stem from what Robert Burns called 'man's inhumanity to man'.

Climate change alarmism is a belief system, and needs to be evaluated as such. There is, indeed, an accepted scientific theory which I do not dispute and which, the alarmists claim, justifies their belief and their alarm. This is the so-called greenhouse effect: the fact that the earth's atmosphere contains so-called greenhouse gases (of which water vapour is overwhelmingly the most important, but CO_2 is another) which, in effect, trap some of the heat we receive from the sun and prevent it from bouncing back into space.

Without the greenhouse effect, the planet would be so cold as to be uninhabitable. But, by burning fossil fuels—coal, oil and gas—we are increasing the amount of CO_2 in the atmosphere and thus, other things being equal, increasing the earth's temperature.

But four questions immediately arise, all of which need to be addressed, coolly and rationally.

First, other things being equal, how much can increased atmospheric CO_2 be expected to warm the earth? (This is known to scientists as climate sensitivity, or sometimes the climate sensitivity of carbon.) This is highly uncertain, not least because clouds have an important role to play, and the science of clouds is little understood. Until recently, the majority opinion among climate scientists had been that clouds greatly amplify the basic greenhouse effect. But there is a significant minority, including some of the most eminent climate scientists, who strongly dispute this.

Second, are other things equal, anyway? We know that over millennia, the temperature of the earth has varied a great deal, long before the arrival of fossil fuels. To take only the past thousand years, a thousand years ago we were benefiting from the so-called medieval warm period, when temperatures are thought to have been at least as warm, if not warmer, than they are today. And during the Baroque era we were grimly suffering the cold of the so-called Little Ice Age, when the Thames frequently froze in winter and substantial ice fairs were held on it, which have been immortalised in contemporary prints.

Third, even if the earth were to warm, so far from this necessarily being a cause for alarm, does it matter? It would, after all, be surprising if the planet were on a happy but precarious temperature knife-edge, from which any change in either direction would be a major disaster. In fact, we know that, if there were to be any future warming (and for the reasons already given, 'if' is correct) there would be both benefits and what the economists call disbenefits. I shall discuss later where the balance might lie.

And fourth, to the extent that there is a problem, what should we, calmly and rationally, do about it?

It is probably best to take the first two questions together.

According to the temperature records kept by the UK Met Office (and other series are much the same), over the past 150 years (that is, from the very beginnings of the Industrial Revolution), mean global temperature has increased by a little under a degree centigrade—according to the Met Office, 0.8°C. This has happened in fits and starts, which are not fully understood. To begin with, to the extent that anyone noticed it, it was seen as a welcome and natural recovery from the rigours of the Little Ice Age. But the great bulk of it—0.5°C out of the 0.8°C—occurred during the last quarter of the twentieth century. It was then that global warming alarmism was born.

But since then, and wholly contrary to the expectations of the overwhelming majority of climate scientists, who confidently predicted that global warming would not merely continue but would accelerate, given the unprecedented growth of global carbon dioxide emissions, as China's coal-based economy has grown by leaps and bounds, there has been no further warming at all. To be precise, the latest report of the Intergovernmental Panel on Climate Change (IPCC), a deeply flawed body whose non-scientist chairman is a committed climate alarmist, reckons that global warming has latterly been occurring at the rate of—wait for it—0.05°Cs per decade, plus or minus 0.1°C. Their figures, not mine. In other words, the observed rate of warming is less than the margin of error.

And that margin of error, it must be said, is implausibly small. After all, calculating mean global temperature from the records of weather stations and maritime observations around the world, of varying quality, is a pretty heroic task in the first place. Not to mention the fact that there is a considerable difference between daytime and night-time temperatures. In any event, to produce a figure accurate to hundredths of a degree is palpably absurd.

The lessons of the unpredicted fifteen-year global temperature standstill (or hiatus as the IPCC calls it) are clear. In the first place, the so-called Integrated Assessment Models which the climate science community uses to predict the global temperature increase which is likely to occur over the next 100 years are almost certainly mistaken, in that climate sensitivity is almost certainly significantly less than they once thought, and thus the models exaggerate the likely temperature rise over the next hundred years.

But the need for a rethink does not stop there. As the noted climate scientist Professor Judith Curry, chair of the School of Earth and Atmospheric Sciences at the Georgia Institute of Technology, recently observed in written testimony to the US Senate:

> Anthropogenic global warming is a proposed theory whose basic mechanism is well understood, but whose magnitude is highly uncertain. The growing evidence that climate models are too sensitive to CO_2 has implications for the attribution of late- twentieth-century warming and projections of 21st-century climate. If the recent warming hiatus is caused by natural variability, then this raises the question as to what extent the warming between 1975 and 2000 can also be explained by natural climate variability.[2]

It is true that most members of the climate science establishment are reluctant to accept this, and argue that the missing heat has, for the time being, gone into the (very cold) ocean depths only to be released later. This is, however, highly conjectural. Assessing the mean global temperature of the ocean depths is—unsurprisingly—even less reliable, by a long way, than the surface temperature record. And in any event most scientists reckon that it will take thousands of years for this 'missing heat' to be released to the surface.

In short, the CO_2 effect on the earth's temperature is probably less than was previously thought, and other things—that is, natural variability and possibly solar influences—are relatively more significant than has hitherto been assumed.

But let us assume that the global temperature hiatus does, at some point, come to an end, and a modest degree of global warming resumes. How much does this matter?

The answer must be that it matters very little. There are plainly both advantages and disadvantages from a warmer temperature, and these will vary from region to region depending to some extent on the existing temperature in the region concerned. And it is helpful in this context that the climate scientists believe that the global warming they expect from increased atmospheric CO_2 will be greatest in the cold polar regions and least in the warm tropical regions, and will be greater at night than in the day, and greater in winter than in summer. Be that as it may, studies have clearly shown that, overall, the warming that the climate models are now predicting for most of this century (I referred to these models earlier, and will come back to them later) is likely to do more good than harm.

This is particularly true in the case of human health, a rather important dimension of wellbeing. It is no accident that, if you look at migration for climate reasons in the world today, it is far easier to find those who choose to move to a warmer climate than those who choose to move to a colder climate. And it is well documented that excessive cold causes far more illnesses and deaths around the world than excessive warmth does.

Pressing down on the alarm button

The 2013-14 IPCC Assessment Report does its best to ramp up the alarmism in a desperate, and almost certainly vain, attempt to scare the governments of the world into concluding a binding global decarbonisation agreement at the crunch UN climate conference due to be held in Paris, 2015. Yet a careful reading of the report shows that the evidence to justify the alarm simply isn't there.

On health, for example, it lamely concludes that 'the world-wide burden of human ill-health from climate change is relatively small compared with effects of other stressors and is not well quantified'[3]—adding that so far as tropical diseases (which preoccupied earlier IPCC reports) are concerned, 'Concerns over large increases in vector-borne diseases such as dengue as a result of rising temperatures are unfounded and unsupported by the scientific literature.' Moreover, the IPCC conspicuously fails to take proper account of what is almost certainly far and away the most important dimension of the health issue. And that is, quite simply, that the biggest health risk in the world today, particularly of course in the developing world, is poverty.

We use fossil fuels not because we love them, or because we are in thrall to the multinational oil companies, but simply because they provide far and away the cheapest source of large-scale energy, and will continue to do so, no doubt not forever, but for the foreseeable future. And using the cheapest source of energy means achieving the fastest practicable rate of economic development, and thus the fastest elimination of poverty in the developing world. In a nutshell, and on balance, global warming is good for you.

The IPCC does its best to contest this by claiming that warming is bad for food production: in its own words, 'negative impacts of climate change on crop yields have been more common than positive impacts'. But not only does it fail to acknowledge that the main negative impact on crop yields has been not climate change but climate change policy, as farmland has been turned over to the production of biofuels rather than food crops. It also understates the net benefit for food production from the warming it expects to occur, in two distinct ways.

In the first place, it explicitly takes no account of any future developments in bioengineering and genetic modification, which are likely to enable farmers to plant drought-resistant crops designed to thrive at warmer temperatures, should these occur. Second, and equally important, it takes no account whatever of another effect of increased atmospheric CO_2, and one which is more certain and better documented than the warming effect. Namely, the stimulus to plant growth: what the scientists call the 'fertilisation effect'. Over the past 30 years or so, the earth has become observably greener, and this has even affected most parts of the Sahel. It is generally agreed that a major contributor to this has been the growth in atmospheric CO_2 from the burning of fossil fuels.

This should not come as a surprise. Biologists have always known that CO_2 is essential for plant growth, and of course without plants there would be very little animal life, and no human life, on the planet. The climate alarmists have done their best to obscure this basic scientific truth by insisting on describing carbon emissions as 'pollution'—which, whether or not they warm the planet, they most certainly are not—and deliberately mislabelling forms of energy which produce these emissions as 'dirty'.

In the same way, they like to label renewable energy as 'clean', seemingly oblivious to the fact that by far the largest source of renewable energy in the world today is biomass, and in particular the burning of dung, which is the major source of indoor pollution in the developing world and is reckoned to cause at least a million deaths a year.

Compared with the likely benefits to both human health and food production from CO_2-induced global warming, the possible disadvantages from, say, a slight increase in either the frequency or the intensity of extreme weather events is very small beer. It is, in fact, still uncertain whether there is any impact on extreme weather events as a result of warming (increased carbon emissions, which have certainly occurred, cannot on their own affect the weather: it is only warming which might). The unusual persistence of heavy rainfall over the UK during February, which led to considerable flooding, is believed by the scientists to have been caused by the wayward behaviour of the jetstream; and there is no credible scientific theory that links this behaviour to the fact that the earth's surface is some 0.8°C warmer than it was 150 years ago.

That has not stopped some climate scientists, such as the publicity-hungry chief scientist at the UK Met Office, Dame Julia Slingo, from telling the media that it is likely that 'climate change' (by which they mean warming) is partly to blame. Usually, however, the climate scientists take refuge in the weasel words that any topical extreme weather event, whatever the extreme weather may be, whether the recent UK rainfall or last year's typhoon in the Philippines, 'is consistent with what we would expect from climate change'.

So what? It is also consistent with the theory that it is a punishment from the Almighty for our sins (the prevailing explanation of extreme weather events throughout most of human history). But that does not mean that there is the slightest truth in it. Indeed, it would be helpful if the climate scientists would tell us what weather pattern would not be consistent with the current climate orthodoxy. If they cannot do so, then we would do well to recall the important insight of Karl Popper— that any theory that is incapable of falsification cannot be considered scientific.

Moreover, as the latest IPCC report makes clear, careful studies have shown that, while extreme weather events such as floods, droughts and tropical storms have always occurred, overall there has been no increase in either their frequency or their severity. That may, of course, be because there has so far been very little global warming indeed: the fear is the possible consequences of what is projected to lie ahead of us. And even in climate science, cause has to precede effect: it is impossible for future warming to affect events in the present.

Of course, it doesn't seem like that. Partly because of sensitivity to the climate change doctrine, and partly simply as a result of the explosion of global communications, we are far more aware of extreme weather events around the world than we used to be. And it is perfectly true that many more people are affected by extreme weather events than ever before. But that is simply because of the great growth in world population: there are many more people around. It is also true, as the insurance companies like to point out, that there has been a great increase in the damage caused by extreme weather events. But that is simply because, just as there are more people around, so there is more property around to be damaged.

The fact remains that the most careful empirical studies show that, so far at least, there has been no perceptible increase, globally, in either the number or the severity of extreme weather events. And, as a happy coda, these studies also show that, thanks to scientific and material progress, there has been a massive reduction, worldwide, in deaths from extreme weather events.

The heavy cost of decarbonisation

It is relevant to note at this point that there is an important distinction between science and scientists. I have the greatest respect for science, whose development has transformed the world for the better. But scientists are no better and no worse than anyone else. There are good scientists and there are bad scientists. Many scientists are outstanding people working long hours to produce important results. They must be frustrated that political activists then turn those results into propaganda. Yet they dare not speak out for fear of losing their funding.

Indeed, a case can be made for the proposition that today's climate science establishment is betraying science itself. During the period justly known as the Enlightenment, science achieved the breakthroughs which have so benefited us all by rejecting the claims of authority—which at that time largely meant the authority of the church—and adopting an overarching scepticism, insisting that our understanding of the external world must be based exclusively on observation and empirical experiment. Yet today all too many climate scientists, in particular in the UK, come close to claiming that they need to be respected as the voice of authority on the subject—the very claim that was once the province of the church.

If I have been critical of the latest IPCC report, let me add that it is many respects a significant improvement on its predecessors. It explicitly concedes, for example, that 'climate change may be beneficial for moderate climate change'—and moderate climate change is all that it expects to see for the rest of this century—and that 'Estimates for the aggregate economic impact of climate change are relatively small ... For most economic sectors, the impact of climate change will be small relative to the impacts of other drivers.' So much for the unique existential planetary threat.

What it conspicuously fails to do, however, is to make any assessment of the unequivocally adverse economic impact of the decarbonisation policy it continues to advocate, which (if implemented) would be far worse than any adverse impact from global warming.

Even here, however, the new report concedes for the first time that the most important response to the threat of climate change must be how mankind has always responded, throughout the ages: namely, intelligent adaptation. Indeed, the 'impacts' section of the latest report is explicitly entitled 'Impacts, Adaptation and Vulnerability'. In previous IPCC reports adaptation was scarcely referred to at all, and then only dismissively.

This leads directly to the last of my four questions. To the extent that there is a problem, what should we, calmly and rationally, do about it?

The answer is—or should be—a no-brainer: adapt. I mentioned earlier that a resumption of global warming, should it occur (and of course it might) would bring both benefits and costs. The sensible course is clearly to pocket the benefits while seeking to minimise the costs. And that is all the more so since the costs, should they arise, will not be anything new: they will merely be the slight exacerbation of problems that have always afflicted mankind.

Like the weather, for example—whether we are talking about rainfall and flooding (or droughts for that matter) in the UK, or hurricanes and typhoons in the tropics. The weather has always varied, and it always will. There have always been extremes, and there always will be. That being so, it clearly makes sense to make ourselves more resilient and robust in the face of extreme weather events, whether or not there is a slight increase in the frequency or severity of such events.

This means measures such as flood defences and sea defences, together with water storage to minimise the adverse effects of drought, in the UK; and better storm warnings, the building of levees, and more robust construction in the tropics.

The same is equally true in the field of health. Tropical diseases—and malaria is frequently (if inaccurately) mentioned in this context—are a mortal menace in much of the developing world. It clearly makes sense to seek to eradicate these diseases—and in the case of malaria (which used to be endemic in Europe) we know perfectly well how to do it—whether or not warming might lead to an increase in the incidence of such diseases.

And the same applies to all the other possible adverse consequences of global warming. Moreover, this makes sense whatever the cause of any future warming, whether it is man-made or natural. Happily, too, as economies grow and technology develops, our ability to adapt successfully to any problems which warming may bring steadily increases.

Yet, astonishingly, this is not the course on which our leaders in the Western world generally, and the UK in particular, have embarked. They have decided that what we must do, at inordinate cost, is prevent the possibility (as they see it) of any further warming by abandoning the use of fossil fuels.

Even if this were attainable—a big 'if', which I will discuss later—there is no way in which this could be remotely cost-effective. The cost to the world economy of moving from relatively cheap and reliable energy to much more expensive and much less reliable forms of energy—the so-called renewables, on which we had to rely before we were liberated by the fossil-fuel-driven Industrial Revolution—far exceeds any conceivable benefit.

It is true that the notorious *Stern Review*—widely promoted by a British prime minister with something of a messiah complex and an undoubted talent for public relations—sought to demonstrate the reverse, and has become a bible for the economically illiterate.

But Stern's dodgy economics have been comprehensively demolished by the most distinguished economists on both sides of the Atlantic. So much so, in fact, that Lord Stern himself has been driven to complain that it is all the fault of the integrated assessment models, which—and I quote him—'come close to assuming directly that the impacts and costs will be modest, and close to excluding the possibility of catastrophic outcomes'.

I suggested earlier that these elaborate models are scarcely worth the computer code they are written in, and certainly the divergence between their predictions and empirical observations has become ever wider. Nevertheless, it is a bit rich for Stern now to complain about them, when they remain the gospel of the climate science establishment in general and of the IPCC in particular.

But Stern is right in this sense: unless you assume that we may be heading for a CO_2-induced planetary catastrophe, for which there is no scientific basis, a policy of decarbonisation cannot possibly make sense.

A similar, if slightly more sophisticated, case for current policies has been put forward by a distinctly better economist than Stern, Harvard's Professor Martin Weitzman, in what he likes to call his 'dismal theorem'. After demolishing Stern's cost-benefit analysis, he concludes that Stern is in fact right but for the wrong reasons. According to Weitzman, this is an area where cost-benefit analysis does not apply. Climate science is highly uncertain, and a catastrophic outcome which might even threaten the continuation of human life on this planet cannot be entirely ruled out however unlikely it may be. It is therefore incumbent on us to do whatever we can, regardless of cost, to prevent this.

This is an extreme case of what is usually termed 'the precautionary principle'. I have often thought that the most important use of the precautionary principle is against the precautionary principle itself, since it can all too readily lead to absurd policy prescriptions. In this case, a moment's reflection would remind us that there are a number of possible catastrophes, many of them less unlikely than that caused by runaway warming, and all of them capable of occurring considerably sooner than the catastrophe feared by Weitzman; and there is no way we can afford the cost of unlimited spending to reduce the likelihood of all of them.

In particular, there is the risk that the earth may enter a new ice age. This was the fear expressed by the well-known astronomer Sir Fred Hoyle in his book *Ice: The Ultimate Human Catastrophe*, and there are several climate scientists today, particularly in Russia, concerned about this. It would be difficult, to say the least, to devote unlimited sums to both cooling and warming the planet at the same time.

At the end of the day, this comes down to judgment. Weitzman is clearly entitled to his, but I doubt it is widely shared; and if the public were aware that it was on this slender basis that the entire case for current policies rested I would be surprised if they would have much support. Rightly so. But there is another problem.

Unlike intelligent adaptation to any warming that might occur, which in any case will mean different things in different regions of the world, and which requires no global agreement, decarbonisation can make no sense whatever in the absence of a global agreement. And there is no chance of any meaningful agreement being concluded. The very limited Kyoto accord of 1997 has come to an end; and although there is the declared intention of concluding a much more ambitious successor, with a UN-sponsored conference in Paris 2015 at which it is planned that this should happen, nothing of any significance is remotely likely.

And the reason is clear. For the developing world, the overriding priority is economic growth: improving the living standards of the people, which means among other things making full use of the cheapest available source of energy: fossil fuels.

The position of China, the largest of all the developing countries and the world's biggest (and fastest growing) emitter of CO_2, is crucial. For very good reasons, there is no way that China is going to accept a binding limitation on its emissions. China has an overwhelmingly coal-based energy sector—indeed it has been building new coal-fired power stations at the rate of one a week—and although it is now rapidly developing its substantial indigenous shale gas resources (another fossil fuel), its renewable energy industry, both wind and solar, is essentially for export to the developed world.

It is true that China is planning to reduce its so-called 'carbon intensity' quite substantially by 2020. But there is a world of difference between the sensible objective of using fossil fuels more efficiently, which is what this means, and the foolish policy of abandoning fossil fuels, which it has no intention of doing. China's total carbon emissions are projected to carry on rising—and rising substantially—as its economy grows.

This puts into perspective the UK's commitment, under the *Climate Change Act*, to near-total decarbonisation. The UK accounts for less than two per cent of global CO_2 emissions: indeed, its total CO_2 emissions are less than the annual increase in China's. Never mind, says Lord Deben, chairman of the government-appointed Climate Change Committee, we are in the business of setting an example to the world.

No doubt this sort of thing goes down well at meetings of the faithful, and enables him and them to feel good. But there is little point in setting an example, at great cost, if no one is going to follow it, and around the world governments are now gradually watering down or even abandoning their decarbonisation ambitions. Indeed, it is even worse than that. Since the UK has abandoned the idea of having an energy policy in favour of having a decarbonisation policy, there is a growing risk that, before very long, our generating capacity will be inadequate to meet our energy needs. If so, we shall be setting an example all right: an example of what not to do.

Challenging the orthodoxy

So how is it that much of the Western world has succumbed to the self-harming collective madness that is climate change orthodoxy? It is difficult to escape the conclusion that climate change orthodoxy has in effect become a substitute religion, attended by all the intolerant zealotry that has so often marred religion in the past, and in some places still does so today.

Throughout the Western world, the two creeds that used to vie for popular support, Christianity and the atheistic belief system of communism, are each clearly in decline. Yet people still feel the need both for the comfort and for the transcendent values that religion can provide. It is the quasi-religion of green alarmism and global salvationism, of which the climate change dogma is the prime example, which has filled the vacuum, with reasoned questioning of its mantras regarded as little short of sacrilege.

The parallel goes deeper. As I mentioned earlier, throughout the ages the weather has been an important part of the religious narrative.

In primitive societies it was customary for extreme weather events to be explained as punishment from the gods for the sins of the people; and there is no shortage of this theme in the Bible, either—particularly, but not exclusively, in the Old Testament. The contemporary version of this is that, as a result of heedless industrialisation within a framework of materialistic capitalism, we have directly (albeit not deliberately) perverted the weather, and will duly receive our comeuppance.

There is another aspect, too, which may account for the appeal of this so-called explanation. Throughout the ages, something deep in man's psyche has made him receptive to apocalyptic warnings that the end of the world is nigh. And almost all of us, whether we like it or not, are imbued with feelings of guilt and a sense of sin. How much less uncomfortable it is, how much more convenient, to divert attention away from our individual sins and reasons to feel guilty, and to sublimate them in collective guilt and collective sin.

Why does this matter? It matters, and matters a great deal, on two quite separate grounds. The first is that it has gone a long way towards ushering in a new age of unreason. It is a cruel irony that, while it was science which, more than anything else, was able by its great achievements, to establish the age of reason, it is all too many climate scientists and their hangers-on who have become the high priests of a new age of unreason.

But what moves me most is that the policies invoked in its name are grossly immoral. We have, in the UK, devised the most blatant transfer of wealth from the poor to the rich—and I am slightly surprised that it is so strongly supported by those who consider themselves to be the tribunes of the people and politically on the left. I refer to our system of heavily subsidising wealthy landlords to have wind farms on their land, so that the poor can be supplied with one of the most expensive forms of electricity known to man.

This is also, of course, inflicting increasing damage on the British economy, to no useful purpose whatever. More serious morally—because it is on a much larger scale—is the perverse intergenerational transfer of wealth implied by orthodox climate change policies. It is not much in dispute that future generations, those yet unborn, will be wealthier than those alive today—ourselves, our children, and for many of us our grandchildren. This is the inevitable consequence of the projected economic growth which, on a 'business as usual' basis, drives the increased carbon emissions which in turn determine the projected future warming. It is surely perverse that those alive today should be told that they must impoverish themselves—by abandoning what is far and away the cheapest source of energy—in order to ensure that those yet to be born, who will in any case be signally better off than they are, will be better off still by escaping the disadvantages of any warming that might occur.

However, the greatest immorality of all concerns the masses in the developing world. It is excellent that, in so many parts of the developing world—the so-called emerging economies—economic growth is now firmly on the march, as they belatedly put in place the sort of economic policy framework that brought prosperity to the Western world. Inevitably, they already account for, and will increasingly account for, the lion's share of global carbon emissions.

But, despite their success, there are still hundreds of millions of people in these countries in dire poverty, suffering all the ills that this brings, in terms of malnutrition, preventable disease, and premature death. Asking these countries to abandon the cheapest available sources of energy is, at the very least, asking them to delay the conquest of malnutrition, to perpetuate the incidence of preventable disease, and to increase the numbers of premature deaths.

Global warming orthodoxy is not merely irrational. It is wicked.

8　Costing climate change

Alan Moran

The IPCC's three voluminous 'Fifth Assessment' Working Group reports and their slimmed-down *Summaries for Policymakers* were completed in 2013 and 2014. To condense the findings of what is said to be the work of 803 authors, the IPCC estimates that a doubling of atmospheric carbon dioxide (CO_2) will cause warming between 1.5°C and 4.5°C. Perhaps in response to seventeen years in which the planet has defied the warming forecasts of climate models, the lower boundary was reduced in the latest assessment. This did not however prevent the decibels of commentary about adverse implications being cranked up.

Highly respected climate scientists put the likely warming below the bottom of the IPCC range. Writing in this volume, Richard Lindzen estimates the maximum warming possible for human induced greenhouse gases is 1°C, while Beenstock, Reingewertz and Paldor find that the relationship between greenhouse gases and warming is spurious except perhaps in the short term.[1]

Entertainers urge us to reduce consumption of non-renewable energy in order to forestall adverse effects of global warming. Ironically some of these exhorters have carbon footprints many times those of common folk—Bono's 2010 world concert tour is estimated to have generated emissions equivalent to the annual level of 6,500 British people.[2]

How much will climate change hurt?

The key questions are, 'How much damage will emanate from the likely atmospheric doubling of carbon dioxide and other greenhouse gases?' and, 'What is the cost of measures to prevent this doubling?' The IPCC puts the qualitative costs of warming in the following foreboding terms:

- Each degree of warming is projected to decrease renewable water resources by at least twenty per cent for an additional seven per cent of the global population.

- Climate change is likely to increase the frequency of droughts.

- Heavy rainfalls are likely to become more intense and frequent.

- In response to further warming by 1°C or more by the mid-twenty-first century and beyond, ocean-wide changes in ecosystem properties are projected to continue, with implications for food security.

- Urban climate change risks, vulnerabilities, and impacts are increasing across the world in urban centres of all sizes, economic conditions, and site characteristics. Climate change will have profound impacts on a broad spectrum of infrastructure systems (water and energy supply, sanitation and drainage, transport and telecommunication), services (including health care and emergency services), the built environment and ecosystem services.

- Climate trends are affecting the abundance and distribution of harvested aquatic species, both freshwater and marine, and aquaculture production systems in different parts of the world but with benefits in other regions.

- Without adaptation, local temperature increases in excess of about 1°C above pre-industrial is projected to have negative effects on yields for the major crops (wheat, rice and maize) with increased global food prices by 2050.

In fleshing out these generalities, the IPCC maintains that climate change is already impacting on natural and human systems. It says these effects include changing precipitation, melting snow and diminishing crop yields.

The Working Group II report sees additional costs being derived from the following key risks:[3]

- Risk of death, injury, ill-health, or disrupted livelihoods in low-lying coastal zones and small island developing states and other small islands, due to storm surges, coastal flooding, and sea-level rise.
- Risk of severe ill-health and disrupted livelihoods for large urban populations due to inland flooding in some regions.
- Systemic risks due to extreme weather events leading to breakdown of infrastructure networks and critical services such as electricity, water supply, and health and emergency services.
- Risk of mortality and morbidity during periods of extreme heat, particularly for vulnerable urban populations and those working outdoors in urban or rural areas.
- Risk of food insecurity and the breakdown of food systems linked to warming, drought, flooding, and precipitation variability and extremes, particularly for poorer populations in urban and rural settings.
- Risk of loss of rural livelihoods and income due to insufficient access to drinking and irrigation water and reduced agricultural productivity, particularly for farmers and pastoralists with minimal capital in semi-arid regions.
- Risk of loss of marine and coastal ecosystems, biodiversity, and the ecosystem goods, functions, and services they provide for coastal livelihoods, especially for fishing communities in the tropics and the Arctic.
- Risk of loss of terrestrial and inland water ecosystems, biodiversity, and the ecosystem goods, functions, and services they provide for livelihoods.

Studies to assess and quantify these concerns include the official ones conducted by large teams of experts. The most prominent of these are the UK report by Nicholas Stern and the Australian report by Ross Garnaut.[4] The latter was followed up by other reports, including the 'Strong Growth Low Pollution' modelling by the Treasury.[5]

As authors, this listed 84 Treasury officials in addition to officers from other agencies.

While arguing that 'the analysis should not focus only on narrow measures of income like GDP', Stern suggested the cost of human induced climate change under business-as-usual would be 'the equivalent of around a twenty per cent reduction in consumption per head, now and into the future'. Combatting this, he said, would cost only one per cent of GDP. Stern received a peerage for the report which Her Majesty's Government has archived.

Like Stern, Garnaut provided a number of cost estimates, including one of up to twelve per cent. He maintained 'all of the detailed assessments of the economics of climate change indicate that the main costs of climate change, and therefore the main benefits of mitigation, accrue in the twenty-second, twenty-third centuries, and beyond.'[6]

Garnaut included many individual features in his warming-induced damage estimates, though the detailed costs were not well supported. Thus, he argues that the additional expense for repairing roads and bridges could cost over one percentage point of GDP but offers no substantiation for these assertions. Oddly, he also argues that other cost increases will ensue from reduced tourism partly due to a highly implausible collapse of the Great Barrier Reef but also because of higher electricity costs and a loss of international tourism (in the reference case international travel to Australia is projected to increase substantially).

Projecting a series of 'climate refugee' scenarios, Garnaut also sees a need for increased defence spending with an additional cost amounting to 0.2 per cent per annum. However, the IPCC has now downgraded such fears. In 2005 the IPCC had global warming creating '50 million "climate refugees" by 2010' (later deferred to 2020). It now says that such fears, 'are not supported by past experiences of responses to droughts and extreme weather events and predictions for future migration flows are tentative at best.'[7]

Richard Tol has pointed out that the Stern and Garnaut reports were not peer reviewed. Tol has now been demonised for withdrawing his name as an author of a key IPCC chapter that he claims the *Summary for Policymakers* had distorted. That Summary argues, 'the incomplete estimates of global annual economic losses for additional temperature increases of 2 degrees Celsius are between 0.2 and 2.0 per cent of income … Losses accelerate with greater warming but few quantitative estimates have been completed for additional warming around 3 degrees Celsius or above.'[8]

The IPCC's estimates of costs from inaction to prevent climate change are, nonetheless, considerably lower than those offered by the British and Australian semi-official government reports.

The IPCC Assessment lists only four studies since 2008 that estimate economic losses due to climate change. One, by Maddison and Rehdanz, is based on 'self-reported happiness' and therefore fails key scientific verifiability tests.[9] Of the others, Nordhaus suggests a loss of GDP of 2.5 per cent with a 3°C warming;[10] a second — by Bosello, Eboli and Pierfederici — puts the GDP loss at 0.5 per cent for a 1.9°C warming;[11] and a third, by Roson and van der Mensbrugghe, estimates a GDP loss of 1.8 per cent for a 2.3°C increase and 4.6 per cent loss for a 4.9°C warming.[12] The studies are summarised in Table 1.

There is considerable water-muddying within the IPCC's *Fifth Assessment Report* about possible scenarios where much higher warming takes place. But policy has to stay grounded with the more plausible possibilities. The world is replete with remote dangers that might just occur and providing for all of these would take up most of global income.

The bottom line is that the cost of global warming that might result from human activities, as reported by the IPCC, is very small. Moreover, the economists estimating these costs have done so on the basis of some highly unreliable evaluations of damage from climate change.

Thus the costs attributed to losses from reduced agricultural output and productivity, rising sea levels, re-allocation of tourist facilities, river

Table 1: Estimates of welfare loss due to climate change

Study	Warming (°C)	Impact (%GDP)	Method	Coverage
Nordhaus (2008)	3.0	-2.5	Enumeration	Agriculture, sea level rise, other market impacts, human health, amenity, biodiversity, catastrophic impacts
Maddison and Rehdanz (2011)	3.2	-11.5	Statistical	Self-reported happiness
Bosello, Eboli and Pierdederici	1.9	-0.5	CGE	Energy demand; tourism; sea level rise; river floods; agriculture; forestry; human health
Roson and van der Mensbrugghe (2012)	2.3 4.9	-1.8 -4.6	CGE	Agriculture, sea level rise, water resources, tourism, energy demand, human health, labor productivity

Source: D. Arent and R. S. J. Tol, "Chapter 11: Key Economic Sectors and Services," Working Group II contribution to the Fifth Assessment Report, Climate Change 2014: Impacts, Adaptation, and Vulnerability, IPCC (Draft, 2014), accessed 17 July 2014, http://ipcc-wg2.gov/AR5/images/uploads/WGIIAR5-Chap10_FGDall.pdf, 74. Table 10-3.

floods and so on are compiled on a static basis. The costs assume people will not modify their behaviours in response to the forecasted gradual changes in temperature, precipitation patterns and tides. Responses to such changes have taken place in the past and should be far more easily accommodated with the more accurate measurements we enjoy today.

In addition, some of the outcomes that the IPCC is projecting are highly dubious. In the IPCC's *Fifth Assessment Report*, there are allegations of desertification of south east Australia's Murray-Darling Basin and of declining yields of major cereals. Both are readily rebutted.

The first stemmed from a politicisation of the process that interpreted drought, the pattern of which is well understood by Australia's more careful scientists, with a permanent change. The drought has broken and the alarmist scientists promoting the theory have been made to look foolish.

The drought costs estimated by the IPCC owe much to the Garnaut report, which compiled greater losses than the IPCC from climate change as part of its narrative. Garnaut put Australia's costs from a 5°C warming by the end of the century at eight per cent of GDP. His report put losses from agriculture at twenty per cent of the total and those losses were predominantly in the Murray-Darling Basin, home to over one-third of Australian farm output and the nation's major irrigation area. According to Garnaut, half of the present day production in the basin would be lost by 2050 and by the end of the century it would no longer support irrigated agriculture. Such projections are total fantasies.

There is no empirical evidence of the Murray Darling turning into the barren region that Garnaut projects. Rainfall measured by Jennifer Marohasy appears in Figure 1.

Figure 1: Murray Darling Basin annual rainfall, 1900-2007

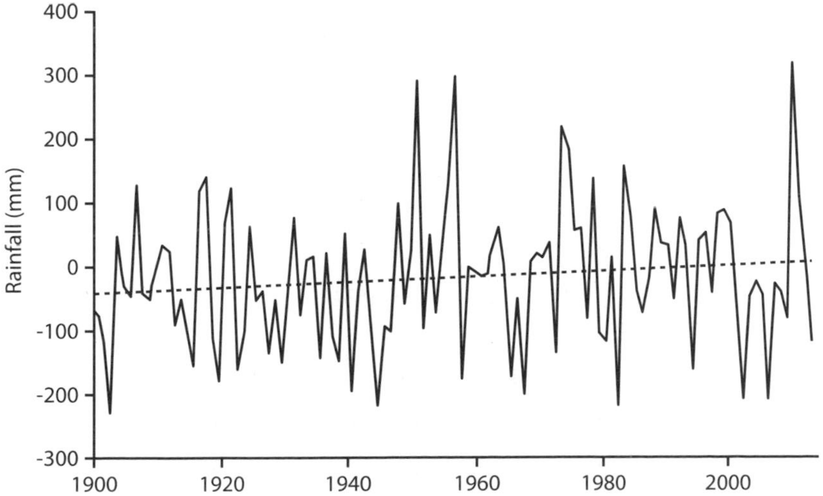

Source: K. Stewart, "IPCC Dud Rainfall Predictions for the Murray-Darling Basin," KensKingdom, 4 April 2014, accessed 17 July 2014, http://kenskingdom.wordpress.com/2014/04/04/ipcc-dud-rainfall-predictions-for-the-murray-darling-basin/.

Drought years in the noughties, as at the turn of the nineteenth century and the 1940s, have been followed by high levels of rainfall and damage through flooding. After half a century of carbon emissions with, according to the IPCC and other warmists, each decade showing higher average temperatures, some corroboration of the impending desertification of south-east Australia should surely be evident. In fact the main threat to irrigated agriculture has come from governments taking water from farmers and allocating it to 'environmental flows.'

The suggestion that yields from major crops are declining as a result of climate change is equally spurious and not supported by empirical evidence.

The following discussion with the IPCC's Chris Field ('one of two lead authors') was broadcast by the Australian Broadcasting Commission:

CHRIS FIELD: Year-on-year, yields have increased by something like two per cent. But they've been increasing by less than that recently, and based on a number of very careful, thorough statistical analyses, researchers are now able to see that for at least two of the world's major food crops, wheat and maize, the increases in yields year-on-year have slowed, partly as a consequence of climate change.

So the drag, the anchoring effect of climate change in making it more and more difficult to increase yields is something we're seeing at the global basis. I'm sure there are some places where there are still yield increases, but there are other places that are offsetting those where yields are decreasing. This idea that we're seeing slower-than-expected yield increases is emerging at the global scale.

SARAH FERGUSON: But we're not just talking about—this isn't any longer about modelling; this is about already-observable facts?

CHRIS FIELD: Absolutely. That's one of the really different things about this report than what the IPCC has said in the past. The impacts of changes that have already occurred are widespread and consequential.[13]

At least with respect to wheat, this is a highly contested view. Exhaustive research by Wilcox and Makowski has found that although higher temperatures mean reduced yields, 'the effects of high CO_2 concentrations (>640 ppm) outweighed the effects of increasing temperature (up to 2 degrees Celsius) and moderate declines in precipitation (up to 20 percent), leading to increasing yields.'[14]

Globally, yields of the major crops have increased at the same rate for decades with no signs of a fall-off. Figure 2 below illustrates global and Australian wheat trends (Australian yields declined for a number of country-specific reasons).

These and other doubtful measurements of human induced climate driven change by the IPCC indicate that its estimate of slender economic loss from climate change is an exaggeration.

Figure 2: Australian and global wheat yield trends

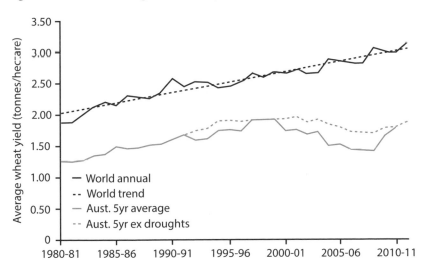

Source: A. Lake, "Australia's declining crop yield trends I: Donald revisited" (paper presented at the 16th Australian Agronomy Conference, University of New England, New South Wales, 14-18 October 2012), accessed July 10, 2014 http://www.regional.org.au/au/asa/2012/agriculture/8163_lakea.htm.

How much will abating climate change hurt?

IPCC Working Group III seeks to assess the costs incurred from stemming human induced greenhouse gas emissions.[15] It does so by quantifying the array of taxes, spending and regulatory measures that it considers necessary. The spin of its *Summary for Policymakers* is evident in its various injunctions, many of which use disturbances claimed from climate change as an agent for income redistribution. The report includes the following points:

- Limiting the effects of climate change is necessary to achieve sustainable development and equity, including poverty eradication.
- Effective mitigation will not be achieved if individual agents advance their own interests independently.
- Issues of equity, justice, and fairness arise with respect to mitigation and adaptation.
- Climate policy intersects with other societal goals creating the possibility of co-benefits or adverse side-effects.

These boilerplate advocacy statements are accompanied by optimistic estimates of the costs which the IPCC has estimated will emanate from the measures it proposes to curtail greenhouse gas emissions.

The highly complex Table 2 summarises modeling of a number of different costs of climate emission reductions and their variations. As is conventionally the case, the modelling is simplified to exclude transitional losses, and the costs of thousands of different taxes and regulatory measures are assumed to be robust and stable.

In the table, to achieve emission concentrations at 450 parts per million in 2050 (row 2), the cost is put at a cumulative 3.4 per cent of world income levels. This cost rises to 4.8 per cent in 2100.

The 2050 cost would increase by 138 per cent (to 4.7 per cent) if carbon capture and storage (CCS) is unavailable.

Table 2: IPCC Estimates of abatement costs

2100 concentration (ppm CO₂eq)	Consumption losses in cost-effective implementation scenarios				Increase in total discounted mitigation costs in scenarios with limited availability of technologies				Increase in mid-and longer term mitigation costs due to delayed additional mitigation up to 2030			
	[% reduction in consumption relative to baseline]			[percentage point reduction in annualised consumption growth rate]	[% increase in total discounted mitigation costs (2015-2100) relative to default technology assumptions]				[% increase in mitigation costs relative to immediate mitigation]			
									$\leq$55 GtCO₂eq		>55 GtCO₂eq	
	2030	2050	2100	2010-2100	No CCS	Nuclear phase out	Limited solar / wind	Limited bio-energy	2030-2050	2050-2100	2030-2050	2050-2100
450 (430-480)	1.7 (1.0-3.7) [N:14]	3.4 (2.1-6.2)	4.8 (2.9-11.4)	0.06 (0.04-0.14)	138 (29-297) [N:4]	7 (4-18) [N:8]	6 (2-29) [N:8]	64 (44-78) [N:8]	28 (14-50) [N:34]	15 (5-59)	44 (2-78) [N:29]	37 (16-82)
500 (480-530)	1.7 (0.6-2.1) [N:32]	2.7 (1.5-42.)	4.7 (2.4-10.6)	0.06 (0.03-0.13)								
550 (530-580)	0.6 (0.2-1.3) [N:46]	1.7 (1.2-3.3)	3.8 (1.2-7.3)	0.04 (0.01-0.09)	39 (18-75) [N:11]	13 (2-23) [N:10]	8 (5-15) [N:10]	18 (4-66) [N:12]	3 (-5-16) [N:14]	4 (-4-11)	4 (-4-11)	16 (5-24)
580-650	0.3 (0-0.9) [N:16]	1.3 (0.5-2.0)	2.3 (1.2-4.4)	0.03 (0.01-0.05)								

Source: IPCC, "Summary for Policymakers," Climate Change 2014: Mitigation of Climate Change. Contribution of Working Group III to the Fifth Assessment Report of the Intergovernmental Panel on Climate Change (Cambridge: Cambridge University Press, 2014), 16.

It would increase a further 7 per cent and 6 per cent respectively if nuclear is phased out, and wind and solar are limited to 20 per cent of energy supply. And it would increase a further 64 per cent if bioenergy fuels are not available, bringing a total of 8.6 per cent by 2100.

With emissions held at 550 parts per million (ppm), baseline costs are stated as 1.7 per cent. Hence, just as the costs of business-as-usual (Table 1) are small, so too are the estimated costs entailed in radically transforming the global economy to achieve the sought after abatement.

The implausibility of this, even with all the yet-to-be-developed technologies, is magnified once the uncertainties of these technologies' performances are factored in. One notable mirage technology is carbon capture and storage (CCS), a program bankrolled in large part by the Australian Government which has recently considerably reduced the funding allocation for the program.

Even if the envisaged new or improved technologies were to be cost-less, the IPCC's estimated losses to the global economy from the forced shift away from fossil fuels are greater than the costs of business-as-usual. Compounding this unfavourable deal is the question of the reliability of the avoidance cost estimates. Economic assessments of minor changes to economies or policy shocks can be modelled with passable degrees of accuracy. But the IPCC modelling attempts to estimate what amounts to a total reorganisation of production, transport and living conditions and to project these a century into the future.

An example of the economic reorganisation entailed is quantified in Figure 3, which shows over the coming fifteen years extraordinary improvements from unknown increases in energy efficiency, a collapse in spending in fossil fuel extraction and a massive reduction in power station investment.

These projections also cover the developing economies which are even more resistant to suicidal economic policies than OECD countries.

Figure 3: Forecasts of future energy investments

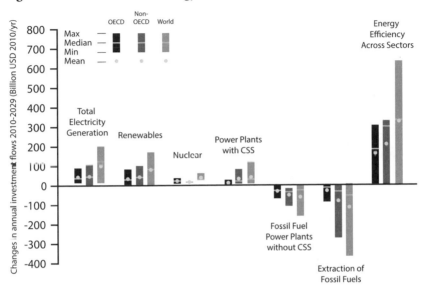

Source: "Summary for Policymakers," Climate Change 2014: Mitigation of Climate Change. Contribution of Working Group III to the Fifth Assessment Report of the Intergovernmental Panel on Climate Change (Cambridge: Cambridge University Press, 2014), 28.

The highly improbable conjecture is that between 2009 and 2029 $550 billion a year extra will be invested in energy efficiency and $540 billion less invested in fossil fuel extraction and fossil fuel plants without CCS. The estimates also suggest over $100 billion a year will be spent on the totally unproven technology of CCS fuel plants with a further $180 billion a year on renewables, which forecasts by the IPCC and official government sources acknowledge will remain three times the cost of coal based electricity generation.

It seems that, having been forced to acknowledge that the much feared global warming has only a trivial effect on real levels of human welfare, the IPCC has to ensure that the estimated costs of its pursuit of the New Jerusalem are not too great. Forcing a radical transformation of society by banning the use of oil and coal and demanding that we reduce energy consumption and shift to horrendously expensive renewables and mythical technology like carbon capture and storage is depicted as a cake walk. Seemingly, only politicians' myopia is standing in the way of a near costless conversion of the global economy away from energy involving high emissions of carbon dioxide.

Just as the government-funded Stern and Garnaut reports produced estimated costs of global warming far in excess of those of the IPCC, these sources also estimated the costs of restraining emissions to be even lower than IPCC expectations. The Stern Report sought reductions in global emissions of CO_2 by 80 per cent of current levels by 2050.[16] Stern argued that the economic cost will be 1 per cent of world GDP, 'which poses little threat to standards of living given that the economic output in the OECD countries is likely to rise by over 200 per cent and in developing countries by more than 400 per cent' during this period.[17]

Neither Stern nor Garnaut has plausibility. Both reports used a near-zero-interest-rates approach to evaluating future costs. Stern used 0.1-1.4 per cent and Garnaut used 1.35-2.65 per cent. A low discount rate means future benefits appear higher than they should be. As the Nongovernmental International Panel on Climate Change argues,

> Discounting is a standard tool of policy analysis on issues ranging from financing public facilities to education and fighting crime. How can climate change be exempted from the use of an analytical tool that is required in all other debates? And if the purpose of reducing greenhouse gas emissions is to benefit future generations, it must be compared to other investments that would do the same thing. Nearly any investment in capital and services that raises productivity and produces wealth benefits future generations. Making investments in emission reductions that yield less than the return on alternative investments in fact impoverishes future generations ... [18]

Figure 4: Price per tonne of carbon dioxide for different emission restraints

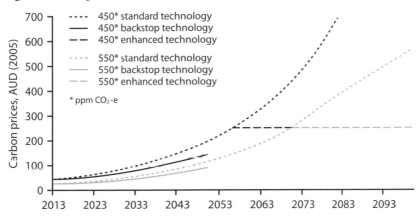

Standard technology refers to the technologies currently in existence

Backstop technology refers to methods that control carbon pricing

Enhanced technology refers to the technologies that do not yet exist

Source: R. Garnaut, *The Garnaut Climate Change Review: Final Report*
(Cambridge: Cambridge University Press, 2008), 251.

The modelling Garnaut commissioned resulted in the prices of CO_2 for the different scenarios shown in Figure 4.

The prices are assumed to level off once they reach $250 per tonne (in 2014 prices about US $285) because it is assumed that technology that is not presently conceptualised will cut in at that price.

Near future carbon prices are estimated to be rather greater than these by the OECD.[19] In US 2007 dollars, the carbon prices necessary for countries or blocs to meet the global goals in 2020 exceed $110 per tonne if each country focused only on its domestic emissions including those from agriculture. If final demand were to be incorporated ('including final demand' in Table 3) the tax would still be $80 per tonne.

The per capita costs of carbon taxes of this magnitude are considerable. For Australia, with emissions of some 18 tonnes per capita (similar to the US and Canada) a tax of $75 per tonne would cost $1350 per

Table 3: Carbon prices in multiple carbon markets scenarios, 2020 (USD 2007/tonne of CO$_2$ equivalent)

Region	Partial	Incl. agriculture	Incl. final demand	Incl. agriculture & final demand	Incl. non-CO$_2$	Incl. non-CO$_2$ & agriculture	Incl. non-CO$_2$ & final demand	All sources
Australia & New Zealand	75	74	60	60	35	20	31	18
Canada	117	112	79	77	57	46	43	36
EU & EFTA	86	83	52	51	28	21	22	17
Japan & Korea	259	257	187	185	178	165	132	124
Other European Annex I countries	21	21	11	11	3	2	3	2
USA	64	59	47	45	26	22	22	19
World*	114	111	81	79	60	52	46	41

*World carbon prices are calculated as an average over acting countries, and weighted by emissions reductions.

Source: E. Landzi, D. Mullaly, J. Chateau and R. Dellink, "Addressing Competitiveness and Carbon Leakage Impacts Arising from Multiple Carbon Markets: A Modelling Assessment," *OECD Environment Working Papers*, No. 58 (2014), 25.

head unless it were to be able to be imposed on all sources, in which case it would be around $250 per head (US and Canada are comparable).[20] These direct costs of a carbon tax understate the true costs since they exclude the costs that are incorporated in the goods and services we buy.

Surveys have indicated that few people, even those voicing concerns about global emissions, are willing to pay these sorts of sums to bring about emission reductions. For Australia, the research firm Galaxy found that only 4 per cent of respondents said they were willing to outlay over $1000 per year to reduce greenhouse emissions (42 per cent said they would outlay over $300 per year).[21]

Of course, saying one would outlay funds and actually outlaying them is different—people, rich and poor, resist tax outlays even if they consider them to be valid.

According to Garnaut, non-hydro renewables and CCS will account for 90 per cent of electricity generation in Australia by 2050. If supply increases by only 20 per cent on the 2014 level of 220,000 GWh, reaching 264,000 GWh, at a current carbon intensity of 0.87 tonnes per MWh, this translates into 230 million tonnes of CO_2.[22] With carbon priced at $250 per tonne, the additional cost of electricity would be $57,500 million, or, for a population which might by then be 30 million, close to an average of $2,000 per capita. And this is only for electricity, which is responsible for less than half of total emissions.

Direct effects on household energy consumers aside, Garnaut trivialises the costs of achieving Australia's required 80 per cent emission reductions, which would necessitate abandoning existing technology and substituting it with totally unproven technologies. The mix of technologies differs from scenario to scenario, but Figure 5 is typical of the mix Garnaut forecasts.

Noteworthy is that by 2050 virtually all electricity is assumed to be generated from technologies that either don't presently exist or are massively more expensive than those of today.

Figure 5: Australia's electricity generation technology shares, 550 ppm scenario

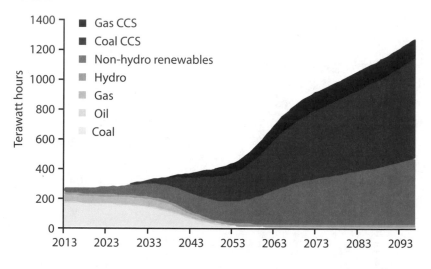

Source: Alan Moran

Critical in estimating the costs is the CO_2 price necessary to drive the changes. This depends on the ease of substituting carbon emitting energy for other forms of energy or replacing energy by other goods and services. If the necessary tax were as low as $100 per tonne, this would treble Australia's wholesale electricity price. Figure 6 illustrates this for Australian electricity supplies.

The only experience of that sort of price shock, which is far less than the IPCC considers necessary, was the quadrupling of oil prices during the 1970s.

However, that brought moderating effects through substitutions from oil to coal and gas and it led to increased oil discoveries. Climate change policy would prevent similar developments.

Figure 6: Tax-inclusive ex-generator electricity costs

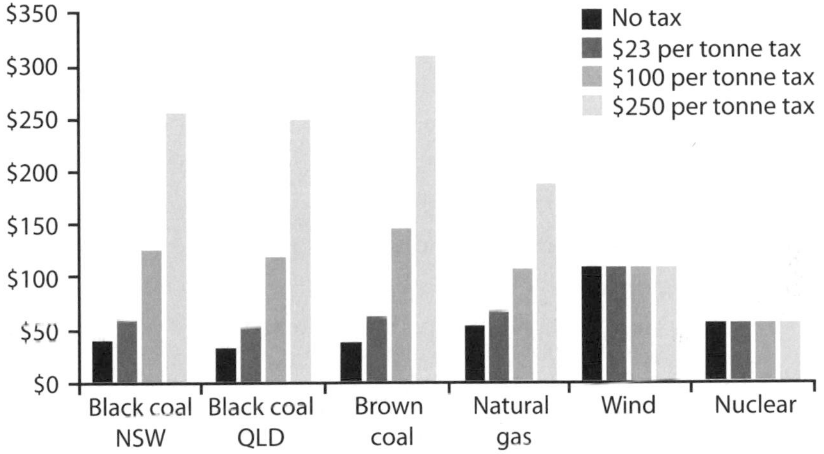

Source: Alan Moran

Conclusion

Just as economic assessments place a relatively low cost of quantifiable damage from climate change, most studies also place a low cost on emission reduction measures. Confidence in such outcomes is clearly not shared by the developing countries which rejected as draconian the measures proposed at Copenhagen in 2009. Nor is the continued resistance of developed countries (with the possible exception of the EU) to take actions involving carbon taxes an indication that there is widespread belief in the low costs promised.

The bottom line is that if global warming is taking place, even the IPCC is forced to acknowledge that it will not be very harmful. According to their own cited studies, the costs are less than a year's annual growth in global GDP. Attempts to suppress emissions of greenhouse gases, even if politically feasible in a multilateral world of nations with different interests, would, on IPCC estimates, cost more than any damage the emissions may be causing. And the costs of such radical action would appear to be grossly understated by the IPCC.

Moreover, the political feasibility of near unanimity of action—without which the abatement assumptions unravel—were shown to be impossible at Copenhagen in 2009 where the increasingly powerful Sino-Indian bloc refused to be persuaded by the threats and blandishments of the EU and its allies. Stand-outs of any significant producer against imposing de facto energy taxes on its businesses would mean that energy-intensive industries will migrate to the lower taxed venue and negate the emission reductions.

Assertions that dire consequences will befall us decades or even centuries into the future make rattling good stories, but when they are unaccompanied by any supporting evidence they have to be treated with caution if they entail high present day costs.

Long range forecasts are fraught with uncertainties and the costs of taking action to obviate a risk must be considered alongside the costs of the risks themselves and the possibilities of taking such action in a world of sovereign states with different interests. And in the case of climate change, the costs as measured are said to be modest.

Even the threefold increase in the costs of energy requires highly optimistic assumptions about low cost replacements for current energy sources. Energy is the most basic of economic resources behind wealth and living standards even though it represents only 5 per cent of GDP (much of which is its distribution costs).

Shifting to the envisaged lower productivity power plants—wind, carbon capture and storage, and solar—means a major reduction in capital productivity, which alongside innovation is the key driver of overall productivity increases.

Finally, the complacency of the IPCC and some other official reports in advocating a near abandonment of current fossil fuels rests on long term forecasts. In addressing the pitfalls of these, one only has to look back to the momentous year of 1914.

A hundred years ago, who would have forecast the fall of the European empires, the rise and fall of communism, the rise of China and India, widespread international air travel, the internet and so on? Back then the few who would have forecast dramatic climate change a century hence would have been proved wrong.

9 Experts as ideologues

James Delingpole

A good many times I have been present at gatherings of people who, by the standards of the traditional culture, are thought highly educated and who have with considerable gusto been expressing their incredulity at the illiteracy of scientists. Once or twice I have been provoked and have asked the company how many of them could describe the Second Law of Thermodynamics. The response was cold: it was also negative. Yet I was asking something which is the scientific equivalent of: Have you read a work of Shakespeare's?

—C.P. Snow[1]

Whenever someone wants to demonstrate beyond all reasonable doubt that an evil, climate-change-denying, kitten-strangling, Big-Oil-funded ignoramus like James Delingpole has nothing useful whatsoever to contribute to the debate on global warming, what they usually ask is this: what are his scientific qualifications?

Well I'll come clean now and admit that mine are close to zero. Sure, I did once grow copper sulphate crystals in a chemistry lesson; yes, I believe I may once have dissected a frog; granted, I do actually possess a grade B in Physics O level. But my university education was in the arts not the sciences. I have an MA in English Language and Literature from

Oxford, which many of the trolls who lurk beneath my climate change blogs seem to think is damning evidence of my unutterable stupidity.

And perhaps they're right. One of the most disturbing things I've noticed during my years in the Climate Wars trenches is the quite terrifying correlation between possession of an English Literature degree and off-the-scale ignorance about the environment.

Take, for example, Roger Harrabin, the BBC environment analyst who so tirelessly bigs up every eco-scare going. Guess what he read at Cambridge.

And Caroline Lucas, Britain's first green MP, who once claimed—in all seriousness—that flying on holiday was as bad as knifing someone in the street because people are 'dying from climate change'.[2] Any idea what her specialist field might have been at university?

And what about Tamsin Omond, the Westminster-educated baronet's granddaughter-turned-hardcore activist who once dumped a truckload of horse poo on Jeremy Clarkson's doorstep in protest at his environmental incorrectness?

Or Bryony—now Baroness—Worthington, the activist from hard-left environmental group Friends of The Earth who was the architect of the most expensive and pointless green legislation in British parliamentary history, the 2008 *Climate Change Act* which commits Britain to spend over £18 billion a year every year till 2050 'decarbonising' its economy.

Or Tim Flannery—who must be a clever fellow: with just a doctorate in kangaroo palaeontology he managed to land a job as Australia's Climate Commissioner which paid A$180,000 a year for a three-day week, all courtesy of the Aussie taxpayer. Go on, have a stab. Your starter for ten: what was Flannery's undergraduate degree at La Trobe University?

Sorry, no prizes. I gave you the answer already—and I hope you find it as puzzling as I do. These people, they're forever invoking the scientific 'consensus' on global warming and saying we should trust the experts at the Intergovernmental Panel On Climate Change and at learned institutions like the Commonwealth Scientific and Industrial Research Organisation and the Royal Society. But what special knowledge are these English Literature graduates bringing to the party that enables them to make these *ex cathedra* pronouncements as to who we should and shouldn't trust? An ability to make a passable, modern language translation of Beowulf? A deep insight into the nature of patriarchy in Georgian England, with special reference to *Emma* and *Pride and Prejudice*? The fact of their having almost understood some bits of James Joyce's *Ulysses*?

All that being said, it's really not my intention here to mock people with English Literature degrees in particular or non-scientists in general. On the contrary, my purpose in this chapter is to demonstrate that they are in some ways better qualified to contribute to the climate change debate than your average scientist. There are two main reasons for this. One is that in the last thirty years or so, the science on climate has become so systematically corrupted that the supposed experts propounding it can no longer be trusted. And the other is that the debate about climate is—and always has been—one which has far more to do with ideology, rhetoric and propaganda than it does with the how-many-angels-can-dance-on-the-head-of-a-pin argument over the extent to which man-made CO_2 emissions may or may not be altering global mean temperatures by fractions of a degree.

But it's amazing how many otherwise well-informed, intelligent people still don't get this. Here, for example, is *Washington Post* columnist Michael Gerson in a piece arguing that the ongoing reluctance of the US public to grapple with global warming stems from an aversion to science. 'The only possible answers come from science. And for non-scientists, this requires a modicum of trust in the scientific enterprise.'[3]

Gerson is speaking here, I suspect, for an awful lot of people who haven't looked too closely into the global warming debate but feel, instinctively, that this must be a scientific issue—to be resolved by scientists—rather than a political one, in which anyone can offer an opinion. Perhaps you're one of them yourself. You'll have read the chapters by say, Ian Plimer, Pat Michaels or Bob Carter and have gone: 'Well that all seems very reasonable. The experts have spoken and the science is perfectly clear. What exactly is the problem?'

The problem, unfortunately, is that scientists like Plimer, Michaels and Carter are rarely heard outside niche publications such as this one or specialist websites like *Watts Up With That?* And this isn't because they're not credible witnesses. Each has scientific credentials as long as your arm but in the crazy world of climate science, knowledge, experience and real world evidence count for less than being on-message. That is, if you support the so-called 'consensus' on man-made global warming then academic tenure, prizes, lavish grants and favourable interviews by Robyn Williams on the ABC are guaranteed; but if you don't, your destiny is to be cast into outer darkness and dismissed as 'anti-science' or a 'denier.'

Consider what happened to Bob Carter when, shortly after the release of the latest IPCC Assessment Report by Working Group I in September 2013, he was interviewed by the BBC's lunchtime radio programme *World At One*. 'Climate has always changed and it always will—there is nothing unusual about the modern magnitudes or rates of change of temperature, of ice volume, of sea level or of extreme weather events,' he said—not unreasonably or inaccurately, for there is a host of data to support every one of these claims.

So how did the climate alarmist establishment respond? As it always does, with a stream of personal abuse. John Ashton, a former 'top climate-change official at the Foreign Office', was given space in *The Guardian* to declare that the decision to give airtime to Carter was 'a betrayal of the editorial professionalism on which the BBC's reputation has been built over generations.'[4] Geneticist Steve Jones—who in 2011 wrote a report for the BBC arguing that it should give less airtime to climate sceptics—said that inviting Carter to give his views represented 'false balance'—and was a bit like inviting a 'homeopath to speak alongside a brain surgeon.' This view was supported at government ministerial level by Greg Barker (a conservative, surprisingly) who said 'I am not trying to ban all dissenting voices but we are doing the public a disservice by treating them as equal, which is not the case.' Bob Ward, of the Grantham Research Institute on Climate Change and the Environment, described it as a 'stunning display of false balance'.

Note the key element missing from all this invective: not once, at any stage, do any of Carter's critics attempt to address the substance of what he said. Why not? Because the facts as he stated them are incontrovertible. Climate *has* always changed. There *have* been periods in the pre-industrial past when temperatures have risen and fallen more rapidly and dramatically from natural causes than anything we have witnessed in this supposed age of catastrophic man-made global warming. The polar ice caps are *not* disappearing. Extreme weather has *always* been with us: every day, somewhere in the world another weather record is being broken because that's what weather does. Sea levels have been rising steadily at the same rate for at least the last 700 years, with no evidence of any alarming recent rise ...

What's going on here, in other words, has nothing whatsoever to do with 'the science.' This is propaganda, pure and simple. Various, apparently trustworthy authority figures have been invoked—the senior diplomat! The politician! The geneticist! The guy from the spiffy-sounding environmental think tank!—to give the impression that when it comes to climate change, all the really serious people in the world are in full agreement. Only a handful of maverick dissenters, the message goes, now dispute the 'consensus.' And only then because they are mad or stupid or in the pay of Big Oil.

Before you dismiss this as a paranoid conspiracy theory, let me give you another example of this climate totalitarianism in action. This one involves Roger Pielke Jr., Professor of Environmental Studies at the University of Colorado, Boulder. Pielke—a self-described 'luke-warmer', that is, a believer in man-made global warming—sits fairly close to the middle of the climate change debate. Yet neither reasonableness nor restraint nor factual accuracy were enough to save his skin when in March 2014 he wrote a piece for the US website *Five ThirtyEight* entitled 'Disasters Cost More Than Ever—But Not Because Of Climate Change'.

The points Pielke made were unexceptionable. A big reinsurance company called Munich Re had published a report claiming the cost of dealing with climate disasters was increasing year on year. Pielke replied, 'When you read that the cost of disasters is increasing, it's tempting to think that it must be because more storms are happening. They're not. All the apocalyptic "climate porn" in your Facebook feed is solely a function of perception. In reality, the numbers reflect more damage from catastrophes because the world is getting wealthier. We're seeing ever-larger losses simply because we have more to lose—when an earthquake or flood occurs, more stuff gets damaged.' And to back up this claim, Pielke cited clear evidence from the most recent IPCC Assessment Report.

For the climate alarmists, however, the truth is no defence. A series of attacks on Pielke's credibility followed: in *The Guardian*; in *Columbia Journalism Review*; in *The Week*; at George Soros's Think Progress; and also, in the comments section below the post, such as this one by 'Top Commenter' Rob Honeycutt, which began: 'Note to Nate Silver ... I'm rather taken aback by this article by Roger Pielke. It's just fundamentally wrong.'

Overwhelmed by the volume of criticism Nate Silver—*Five ThirtyEight's* editor-in-chief—decided to apologise for what Pielke had written and to commission a rebuttal. You do wonder, though, whether he would have caved quite so easily, had he been aware of this comment on a climate alarmist website called *Skeptical Science*, written in February 2011. 'I think this is a highly effective method of dealing with various blogs and online articles where these discussions pop up. Flag them, discuss them and then send in the troops to hammer down what are usually just a couple of very vocal people. It seems like lots of us are doing similar work, cruising comments sections online looking for disinformation to crush. I spend hours every day doing exactly this. If we can coordinate better and grow the "team of crushers" then we could address all the anti-science much more effectively.'

The author of that comment was one Rob Honeycutt. (If the name sounds vaguely familiar, just remind yourself who led the assault on Pielke's article at *FiveThirtyEight*). What he's advocating here is a technique known in the trade as 'astroturfing', where a relatively small number of committed activists are able to punch above their weight by giving the impression of being an extensive, grassroots movement.

Anyone who has ever spoken up publicly for the climate sceptic cause will be painfully familiar with these dirty trick tactics—not just astroturfing, but much uglier stuff like identify theft and false accusations, bullying and attempts at professional assassination. Some—such as the distinguished Swedish meteorologist Lennart Bengtsson—find the experience so dispiriting they never do dare so again.

In May 2014, Professor Bengtsson announced that he was to join the advisory board of the Global Warming Policy Foundation, a sceptical think tank founded by former British Chancellor of the Exchequer Lord Lawson. It was a brave and principled move for Bengtsson who had spent most of his long career, specialising in climate modelling, in the warmist camp. What changed his mind was the evidence: recognising the increasing divergence between the alarmists' climate models and the real world data, he realised that the 'consensus' was flawed and that he could no longer support it.

Just three weeks after taking the position, however, Bengtsson was forced to resign. He had been so badly bullied by his former colleagues— he wrote in his resignation letter—that his health was suffering and he was unable to work. 'It is a situation that reminds me about the time of McCarthy. I would never have expected anything similar in such an original peaceful community as meteorology.'[5]

There is something rather charming about the innocence of that last statement. And it's a view, no doubt, that many people will fondly share—this idea that meteorologists, indeed scientists generally, are a breed apart from the rest of humanity. Unlike, say, bankers or international footballers or businessmen, they are motivated not by vulgar concerns such as money or power but purely by their disinterested quest for knowledge.

One of the people responsible for this popular notion of scientists as a breed apart was the man I quoted at the beginning, C.P. Snow. Since his influential 1959 lecture that myth has grown and grown, assiduously promoted by organisations like the BBC, which rarely lets a day go by without some new paean to the brilliance of the elect white-lab-coated ones with their magical PhDs and their insights into the mysteries of the world which mere mortals cannot ken. Think how many newspaper articles begin with the phrase 'Scientists say … '—the implicit assumption being that whatever these guys say must be true because, well, they're *scientists*.

Now I certainly wouldn't wish to suggest that all scientists are frauds. But I do think that this modern reverence for the profession is not just overdone but also potentially rather dangerous. Essentially, it's a version of the rhetorical fallacy known as the *argumentum ad verecundiam*—the appeal to authority. No person or institution is infallible: neither a Nobel-prize winning geneticist like Sir Paul Nurse, nor yet a scientific body as distinguished as the one of which he is president, the Royal Society (founded 1660). Indeed, it was in acknowledgement of this truth that for the first two centuries of the Royal Society's existence, its house journal *Philosophical Transactions* carried the following 'Advertisement': ' … It is an established rule of the Society, to which they will always adhere, never to give their opinion, as a Body, upon any subject, either of Nature or Art, that comes before them.'

The reason for this was quite simple: scientific knowledge is not fixed. If it were—if all nature's secrets were known—what would be the point of being a scientist? But what ought to be immediately obvious to anyone who ponders the logic is apparently anathema to the current climate establishment. From NASA GISS in the US to the Climatic Research Unit at the University of East Anglia, from the Royal Society to the CSIRO, from the IPCC to the UK Met Office, the official message is the same: the 'science' on climate is 'settled.' Only a handful of fruit-loops, heretics and right-wing ideologues now dispute the 'consensus' on global warming.

This is not 'science' we're seeing in action, here, but a form of po-litical activism. One high-profile climate alarmist, the late Stephen Schneider—an IPCC lead-author and professor of Environmental Biology and Global Change at Stanford University—came close to ad-mitting this in an interview with *Discover* magazine. ' ... Like most people we'd like to see the world a better place, which in this context translates into our working to reduce the risk of potentially disastrous climate change ... That of course entails getting loads of media coverage. So we have to offer up scary scenarios, make simplified, overdramatic statements and make little mention of any doubts we might have.'[6]

Charles Darwin would have disagreed. 'A scientific man ought to have no wishes, no affections ... a mere heart of stone,' he once said, and of course he is right. True science is about discovering what 'is' not what 'ought to be'. It's about the rigorous application of the 'scientific method', which values free, open enquiry, embraces dissent and stands or falls on empirical observation. This means, for example, that when a once-plausible theory is 'falsified' by real-world data, that theory be-comes bunk—no matter what all the learned scientific institutions may claim to the contrary.

Once you appreciate that much of what is going on in climate science is quintessentially unscientific, the viciousness of the debate makes a lot more sense. The 'science' is out there and has been for some time: Anthropogenic Global Warming theory is a busted flush. But instead of conceding the point to the opposition, an arrogant, dishonest, ruthless climate establishment continues to prop up its outmoded hypothesis by increasingly nefarious means.

Like most of those who have engaged on the behalf of the sceptical cause, I have bitter personal experience of these methods. A few years ago, a BBC producer wrote politely to ask whether I would appear in a documentary on global warming. The presenter, she promised me, had a completely open mind on the subject and was anxious to consult my opinion because I was one of the journalists who had exposed the Climategate scandal. I happily consented, not least because the presenter was an impressive-sounding Nobel-prize-winning future head of the Royal Society called Sir Paul Nurse.

Yes, with hindsight I should have known better. It was, of course, intended all along as a hatchet-job designed to discredit the cause of climate scepticism. Sceptics were bracketed in the same category as opponents to GM crops and people who don't believe there's a connection between HIV and AIDs. And about the only section of the gruelling four-hour interview with me they used was one in which Nurse tried to catch me out with a weird analogy. Being against the 'consensus' on climate change, Nurse suggested, is the same as if you've got cancer and instead of seeking treatment through mainstream medicine you choose some alternative therapy quack cure.

As a professional writer and polemicist I have no problem in principle with analogies. I often use them myself when discussing the Climate Wars. For example, I have sometimes argued that the warmist establishment's desperate attempts to prop up its fatally flawed hypothesis is as absurd as if, shortly after news began to break about the sinking of the

Titanic, the ship's owners the White Star Line had held a press conference insisting that, no, the Titanic was continuing to steam ahead to New York.

The analogy works because that's pretty much exactly what is happening right now with mainstream climate science. All the models say one thing: the Titanic is unsinkable; the ship is sailing at so many knots on such-and-such a course and therefore will arrive at its destination on the scheduled date. But observed reality says something rather different —and no amount of cunning spin, or establishment rank-closing, or legal threats can ever possibly make it otherwise.

Nurse's analogy does not work, however, because this 'cancer' he and his fellow warmists have diagnosed is starting to look like nothing more serious than an ingrown toenail. Do we really want to go through the misery of chemotherapy or surgery on the precautionary principle that at some unspecified future date the toenail problem might suddenly metastasise into something deadlier? That's certainly where we are at the moment with regard to international policy on climate change: billions if not trillions of dollars are being diverted into renewable energy schemes, decarbonisation projects, biofuels and such like for no better reason than that, some while back, a bunch of scientists came up with a theory that anthropogenic CO_2 posed a serious health risk to the planet.

But since then, the science has moved on—just like it did with humours, and phlogiston, and eugenics, and all those other theories that were fashionable for a period but which were eventually superseded by what Thomas Kuhn called a 'paradigm shift' in scientific thinking. This is how true science works and how it always has worked. A scientific theory, Karl Popper argued, is only useful if it contains the key to its own destruction. That is, for a proposition to have any real scientific value it must be 'falsifiable'—capable of being proved wrong through experiment and observation. Just like has happened to the once-plausible, but now thoroughly discredited theory of catastrophic anthropogenic global warming.

You'd think a scientist of the calibre of Sir Paul Nurse—or indeed of Michael Mann, director of the Earth System Science Center at Penn State University, or Phil Jones, Director of the Climatic Research Unit at the University of East Anglia—would be familiar with this kind of thing. But apparently not. Thank God for know-nothing English Literature graduates and their pesky insistence on doing a bit of background reading, eh?

10 Uncertainty, scepticism and the climate issue

Garth W. Paltridge

The World Meteorological Organization of the United Nations took its first steps towards establishing the World Climate Program in the early 1970s.[1] Among other things, it held an international workshop in Stockholm to define the main scientific problems which needed to be solved before reliable climate forecasting could be possible. The workshop defined quite a number of problems, but focused on just two.

The first concerned an inability to simulate the amount and character of clouds in the atmosphere. Clouds are important because they govern the balance between solar heating and infrared cooling of the planet, and thereby are a control of Earth's temperature. The second concerned an inability to forecast the behaviour of oceans. Oceans are important because they are the main reservoirs of heat in the climate system. They have internal, more-or-less random, fluctuations on all sorts of time-scales ranging from years through to centuries. These fluctuations cause changes in ocean surface temperature that in turn affect Earth's overall climate.

The situation hasn't changed all that much in the decades since. Many of the problems of simulating the behaviour of clouds and oceans are still there (along with lots of other problems of lesser moment) and for many of the same reasons as were appreciated at the time. Perhaps the most significant is that climate models must do their calculations at each point of an imaginary grid of points spread evenly around the world at various heights in the atmosphere and depths in the ocean. The calculations are done every hour or so of model time as the model steps forward into its theoretical future. Problems arise because practical constraints on the size of computers ensure that the horizontal distance between model grid-points may be as much as a degree or two of latitude or longitude—that is to say, a distance of many tens of kilometres.

That sort of distance is much larger than the size of a typical piece of cloud. As a consequence, simulation of clouds requires a fair amount of inspired guesswork as to what might be a suitable average of whatever is going on between the grid-points of the model. Even if experimental observations suggest that the models get the averages roughly right for a short-term forecast, there is no guarantee they will get them right for atmospheric conditions several decades into the future. Among other problems, small errors in the numerical modelling of complex processes have a nasty habit of accumulating with time.

Again because of this grid-point business, oceanic fluctuations and eddies smaller than the distance between the grid-points of a model are unknown to that model. This would not be a problem except that eddies in turbulent fluids can grow larger and larger. A small random eddy in the real ocean can grow and appear out of nowhere as far as a forecasting model is concerned, and make something of a dog's breakfast of the forecast from that time on.

All of the above is background to one of the great mysteries of the climate change issue. Virtually all the scientists directly involved in climate prediction are aware of the enormous problems and uncertainties still associated with their product. How then is it that those of them involved in the latest report of the Intergovernmental Panel on Climate Change (the IPCC) can put their hands on their hearts and maintain there is a 95 per cent probability that human emissions of carbon dioxide (CO_2)

have caused most of the global warming that has occurred over the last several decades?

Bear in mind that the representation of clouds in climate models (and of the water vapour which is intimately involved with cloud formation) is such as to amplify the forecast warming from increasing atmospheric CO_2—on average over most of the models—by a factor of about three. In other words, two-thirds of the forecast rise in temperature derives from this particular model characteristic. Despite what the models are telling us—and perhaps *because* it is models that are telling us—no scientist close to the problem and in his right mind, when asked the specific question, would say that he is 95 per cent sure that the effect of clouds is to amplify rather than to reduce the warming effect of increasing CO_2. If he is not sure that clouds amplify global warming, he cannot be sure that most of the global warming is a result of increasing CO_2.

Bear in mind too that no scientist close to the problem and in his right mind, when asked the specific question, would say there is only a very small possibility (that is, less than 5 per cent) that internal ocean behaviour could be a major cause of the warming over the past half-century. He would be particularly careful not to make such a statement now that there has been no significant warming over the most recent fifteen-or-so years. In the mad scurry to find reasons for the pause, and to find reasons for an obvious failure of the models to simulate the pause, suddenly we are hearing that perhaps the heat of global warming is being 'hidden' in the deep ocean. In other words we are being told that some internal oceanic fluctuation may have reduced the upward trend in global temperature. It is therefore more than a little strange that we are not hearing from the IPCC (or at any rate not hearing very loudly) that some natural internal fluctuation of the system may have given rise to most of the earlier upward trend.

In the light of all this, we have at least to consider the possibility that the scientific establishment behind the global warming issue has been drawn into the trap of seriously overstating the climate problem—or, what is much the same thing, of seriously understating the uncertainties associated with the climate problem—in its effort to promote the cause. It is a particularly nasty trap in the context of science, because it risks destroying, perhaps for centuries to come, the unique and hard-won reputation for honesty which is the basis of society's respect for scientific endeavour. Trading reputational capital for short-term political gain isn't the most sensible way of going about things.

The trap was set in the late 1970s when the environmental movement first realised that doing something about global warming would play to quite a number of its social agendas. At much the same time, it became accepted wisdom around the corridors of power that government-funded scientists (that is, most scientists) should be required to obtain a goodly fraction of their funds and salaries from external sources—external anyway to their own particular organisation.

The scientists in environmental research laboratories, since they are not normally linked to any particular private industry, were forced to seek funds from other government departments. In turn this forced them to accept the need for advocacy and for the manipulation of public opinion. For that sort of activity, an arms-length association with the environmental movement would be a union made in heaven. Among other things it would provide a means by which scientists could distance themselves from responsibility for any public overstatement of the significance of their particular research problem.

The trap was partially sprung in climate research when a significant number of the relevant scientists began to enjoy the advocacy business. The enjoyment was based on a considerable increase in funding and employment opportunity. The increase was not so much on the hard-science side of things but rather in the emerging fringe institutes and organisations devoted, at least in part, to selling the message of climatic doom. A new and rewarding research lifestyle emerged which involved the giving of advice to all types and levels of government, the broadcasting of unchallengeable opinion to the general public, and easy justification for attendance at international conferences—this last in some luxury by normal scientific experience, and at a frequency previously unheard of.

Somewhere along the line it came to be believed by many of the public, and indeed by many of the scientists themselves, that climate researchers were the equivalent of knights on white steeds fighting a great battle against the forces of evil—evil, among other things, in the shape of 'big oil' and its supposedly unlimited money. The delusion was more than a little attractive.

The trap was fully sprung when many of the world's major national academies of science (the Royal Society in the UK, the National Academy of Sciences in the US, and the Australian Academy of Science) persuaded themselves to issue reports giving support to the conclusions of the IPCC. The reports were touted as national assessments that were supposedly independent of the IPCC and of each other, but of necessity were compiled with the assistance of, and in some cases at the behest of, many of the scientists involved in the IPCC international machinations. In effect, the academies, which are the most prestigious of the institutions of science, formally nailed their colours to the mast of the politically correct.

Since that time in 2010-11 or thereabouts, there has been no comfortable way for the scientific community to raise the spectre of serious uncertainty about the forecasts of climatic disaster. It can no longer use the environmental movement as a scapegoat if it should turn out that the threat of global warming has no real substance. It can no longer escape prime responsibility if it should turn out in the end that doing something in the name of mitigation of global warming is the costliest scientific mistake ever visited on humanity. The current re-direction of global funds in the name of climate change is of the order of a billion dollars a day. And in the future, to quote US Senator Everett Dirksen, 'a billion here and a billion there, and pretty soon we'll be talking about real money.'

At the same time, the average man in the street, a sensible chap who these days can smell the signs of an oversold environmental campaign from miles away, is beginning to suspect that it is politics rather than science which is driving the issue.

Scientists—most scientists anyway—may be a bit naïve, but they are not generally wicked, idiotic, or easily suborned either by money or by the politically correct. So whatever might be the enjoyment factor associated with supporting officially accepted wisdom, and whatever might be the constraints applied by the scientific powers-that-be, it is still surprising that the latest IPCC report has been tabled with almost no murmur of discontent from the lower levels of the research establishment. What has happened to the scepticism that is supposedly the lifeblood of scientific enquiry?

The answer probably gets back to the uncertainty of it all. The chances of proving—proving in the hard scientific sense—that change of climate over the next century will be large enough to be disastrous are virtually nil. The same uncertainty ensures that the chances of a climate sceptic, or anyone else for that matter, proving the disaster theory to be oversold are also virtually nil. To that extent there is a level playing field for the two sides of the argument. The problem is that climate research necessarily involves enormous resources, and is a game for institutions and organisations. Scepticism is an occupation for individuals. Things being as they are in the climate change arena, scepticism by an individual within the system can be fairly career limiting. In any event, most individual scientists have a conscience, and are reluctant to put their head above the public parapet in order to propound a view of things that is highly uncertain and may indeed be inherently unprovable.

There is a broader context to this issue of uncertainty.

To the extent that there is such a thing as normal science, it relies upon accurate observations to verify its theories. 'Accurate' is the operative word here. Climate research has to rely on spectacularly inaccurate data for information on Earth's past climate. Even though there are vast amounts of atmospheric and oceanographic data to play with, together with lots of proxy information from tree rings and ice cores and corals and so on, abstracting a coherent story from it all is something of a statistical nightmare. It gives a whole new meaning to the old saying popularised by Mark Twain about 'lies, damn lies and statistics.'

Suffice it to say that climate science is an example of what Canadian educator Sue McGregor calls 'post-normal science' in which 'the facts are uncertain, values are in dispute, stakes are high and decisions are urgent.' In such circumstances it is virtually impossible to avoid sub-conscious cherry-picking of data to suit the popular theory of the time. Even Isaac Newton and Albert Einstein were not immune from the problem. In their case they were of sufficient genius (and were sufficiently lucky!) for their theories ultimately to trump the inaccuracy of the observations they had selected. Other scientists are rarely so prescient or so lucky. In the modern era of concern about climate, the problem is compounded by the existence of vastly complex computer forecasting models that can be tuned, again more-or-less subconsciously, to yield a desired result. From theory to observation and back again—if we are not very careful, the cherry-picking can go round and round in an endless misleading loop.

But the real worry with climate research is that it is on the very edge of what is called postmodern (as opposed to post-normal) science. Postmodern science is a counterpart of the relativist world of postmodern art and design. It is a much more dangerous beast where results are valid only in the context of society's beliefs, and where the very existence of scientific truth can be denied. Postmodern science envisages a sort of political nirvana in which scientific theory and results can be consciously and legitimately manipulated to suit either the dictates of political correctness or the policies of the government of the day.

There is little doubt that some players in the climate game—not a lot, but enough to have severely damaged the reputation of climate scientists in general—have stepped across the boundary into postmodern science. The *Climategate* scandal of 2009 for instance, wherein thousands of emails were leaked (or perhaps hacked) from the Climate Research Unit of the University of East Anglia, showed that certain senior members of the research community were, and presumably still are, quite capable of deliberately selecting data in order to overstate the case for dangerous climate change. The emails showed as well that these senior members were quite happy to discuss ways and means of controlling the research journals so as to deny publication of any material that goes against the orthodox dogma. The ways and means included the sacking of recalcitrant editors.

Whatever the reason, it is indeed vastly more difficult to publish results in climate research journals if they run against the tide of politically correct opinion, which is why most of the sceptic literature on the subject has been forced onto the web, and particularly onto web-logs devoted to the sceptic view of things. This, in turn, is why the more fanatical believers in disastrous anthropogenic global warming insist that only peer-reviewed literature should be accepted as an indication of the real state of affairs. They argue that the sceptic blogs should never be taken seriously by 'real' scientists, and certainly should never be quoted.

This is a great pity. Some of the sceptics are extremely productive as far as critical analysis of climate science is concerned. Names like Judith Curry (Chair of the School of Earth and Atmospheric Sciences at the Georgia Institute of Technology), Steve McIntyre (a Canadian geologist-statistician), and blogger Willis Eschenbach come to mind. These three in particular provide a balance and maturity in public discussion that puts many players in the global warming movement to shame, and as a consequence their outreach to the scientifically-inclined general public is highly effective. Their output, together with that of other sceptics on the web, is well on the way to becoming a practical and stringent substitute for peer review.

Once upon a time we were led to believe that the road to fame and fortune within science was to produce new ideas that challenged accepted belief. Preferably, those new ideas would lead to tangible benefits for society. But irrespective of the benefit side of things, the practical basis of all research was to be openly sceptical about everything—particularly about one's own theories, and particularly about any new theory that had some vague connection to politically correct ideas of the day. Conscious, deliberate and obvious scepticism was regarded as essential to maintaining some sort of immunity from the human failing of seeing what one wants to see rather than what is real. Good scientific practice demanded at the very least that one should present the evidence against a new theory at the same time as the evidence for it.

It seems that science is not what it used to be. In those parts of it that bear upon the politically correct, sceptics are frowned upon, given nasty names, and ultimately may have their reputations burned at the stake. Certainly in the field of climate change, one could perhaps be forgiven for thinking that advocacy for the cause trumps the need for scepticism on any day of the week. This is no small problem in the grand scheme of things, because the whole issue of climate change has lots to be sceptical about.

The take-home message is that there is more than enough uncertainty associated with forecasting climate to allow normal human beings to be reasonably hopeful that global warming might not be as bad as is currently touted. Climate scientists, and indeed scientists in general, are not so lucky. Largely as a consequence of their decision to insulate themselves from sceptical opinion, they have a lot to lose if time should prove them wrong.

11 The trillion dollar guess and the zombie theory

Jo Nova

Scientifically, the theory of a carbon disaster started knocking on death's door ten years ago. It quietly went 'terminal' but hardly anyone knew. Over the last decade 28 million weather balloons, 30 years of satellite recordings and 3,000 robotic ocean buoys confirmed that if the carbon disaster wasn't dead, it was on the critical list—not critically important, but critically wounded. Few realised that a trillion dollar industry was based entirely on a guess made in 1896 about relative humidity, and that the guess appears to be wrong.

The first climate advisory committee (namely the compilers of the 'Charney report', convened by the National Academy of Sciences as a kind of baby IPCC) repeated the assumption in 1979[1], and over the next four decades Western governments would commit to a grand project to try to change the weather. At one point global carbon markets reached a turnover of US $176 billion a year.[2] The Bank of America pledged $50 billion to combat climate change (how green is your banker?).[3] Renewables investment reached $359 billion annually.[4] Such was the cult-like fear, the EU unconvincingly boasted that they had agreed to 'commit at least 20% of EU spending in the period 2014-2020' to 'climate action objectives.'[5] What was scary was that nobody laughed.

Most of this money depended on an assumption made about relative humidity in the upper troposphere. Like triple A-rated mortgages, the real uncertainty was written in fine print while the theory was advertised as 'simple physics'.

The guess that created the trillion dollar crisis

It seemed like a good idea at the time. Water vapour (aka humidity) is a more powerful greenhouse gas than CO_2. Warmer air can hold more water vapour. What if CO_2 warmed the world, which caused humidity to rise and amplified the warming?[6] Catastrophe.

But water molecules are the starring split-personality-molecules of chemistry. Humidity has options: in an instant it can be cloud, rain, ice or snow, and all of these have different impacts on the climate. Humidity warms the Earth, but most clouds cool it. The extra warming caused by CO_2 could be amplified or undone by what the fickle water molecules do.

Water is the real dynamo controlling the climate. While CO_2 is steadily increasing year on year, levels of water in the atmosphere change by the hour. A CO_2 molecule might float for years, but each water molecule stays aloft for only ten days or so, and once it has become part of a cloud, it condenses out in less than a day. The amplification is called positive feedback, and this particular feedback from water molecules is one of the biggest single factors in climate models.[7] There are claims that it doubles the effect of all other forms of warming.[8]

Here's the line from the 1979 Charney report:

> A plausible assumption, borne out quantitatively by model studies, is that relative humidity remains unchanged. The ... increase in absolute humidity ... provides a positive feedback.[9]

That's it: the foundation for multinational global action comes from a 'plausible assumption'. Hey—but it was backed by 1970s computer models. They go on to say those same models 'assume fixed relative humidity'. What's plausible is that if you use models that assume relative humidity stays the same, those models will confirm that relative humidity will stay the same.

Today, not much has changed. The modelers assume that CO_2 has caused most of the warming since the industrial revolution, and for the most part, they also assume relative humidity stays the same. The models then show that CO_2 caused most of the warming and thus the assumption about humidity was 'right'. If some other factor caused some of that warming both points would be wrong.

The hot-spot that wasn't

The humidity that supposedly amplifies the warming is not just any old humid patch anywhere, but the thin layer near the top of the troposphere, about ten to twelve kilometres above the tropics. This is where the action is. Models predict faster warming there, and the trends show up as a red 'hot-spot' on graphs. But it's hard to measure. It's not like scientists can poke gauges up there on long sticks from the office.

The best data we have comes from weather balloons, which rise up through the layer and radio the information back before they explode. We've released 28 million or so of these since the late 1950s, and the trend up there is unmistakably not what the models expected. Instead of getting more humid as the air warmed, it got less.[10] Temperatures also didn't warm as much as they were supposed to.[11] The result was stark in the colour maps of the atmosphere. Yellow is not red.

Where was that positive feedback?

Other data kept coming in too. Temperature recordings from all four major global databases unexpectedly flat-lined together, which wasn't a problem itself, except that it showed that the models don't understand the climate. Around the world, 6,000 boreholes were drilled in ocean mud,[12] and rocks, stalagmites, corals and clamshells were used to estimate the last 2,000 years of temperature.[13] The message was clear: there were global ups and downs that have nothing to do with CO_2. Some mystery factor is moving temperatures on Earth and the models don't know what it is, and it's more important than CO_2. Perhaps that mystery factor is working now, perhaps it isn't. The models can't tell us.

In response, the scientific-financial-green complex swung into action. Postmodern Dadaist scientific papers appeared. Everyone from UN committees to investment bankers churned out glossy reports (which everyone cited but almost no one actually read). Vice Presidents did full-fear documentaries and black belt graphs evolved to protect the dead. Fans cheered, and blind journalists applauded. *Let no man ask a difficult question!* It's really been a spectacular public relations effort.

Can I sell you a used theory?

There's a shell game going on with evidence. Almost all of the pin-ups of climate change are irrelevant because there's no cause and effect link. It's true the world is warming, sea levels are rising, glaciers are melting, and small fish are getting reckless.[14] But the *effects* of all the causes of warming are largely the same. Whether it is the sun, cosmic rays or a Klingon plot, seas would rise, glaciers would melt, and heatwaves roll on. The real problem, then—the $2 trillion question—is how to tell 'wot did it' and *by how much*. It's a multivariable nightmare. All factors are changing simultaneously, and there are no controls, and no reruns. Strip back the advertised 'signs of warming', and the sacred-vault-of-95-per-cent-certainty contains almost nothing pointing the finger at CO_2. The climate simulations are 'it'.

All the talk of 'it' being 'simple physics' is, and always was, a complete red flag. Two-thirds of the forecast of doom comes from complex, debatable feedbacks, not the simple physics of CO_2.

The models are consistent. They're bad at everything.

With a bad assumption at the core, it's no wonder the models don't work. 98 per cent of the models predicted that there were no circumstances where global surface temperatures would pause for as long as fifteen years.[15] The pause has now been somewhere around seventeen years long, or twenty depending on who is counting.[16] As *The HockeySchtick* blog says, 'If you can't explain the "pause", you can't explain the cause'.[17]

Thus and verily the excuses for the pause have flowed: humans were once forecast to slow the winds,[18] but then faster winds arrived instead, and so now, perchance, they could have caused the pause.[19] Likewise, volcanoes could have *cooled* the Earth lately with aerosol particulates[20] (although peak volcanic aerosols were higher in the 1980s and 1990s.[21]) A new force called 'natural variability' has been invoked too. But no one can quite explain why nature only cools the world and never causes the warming.

Likewise, it's also possible that the missing heat *could* have gone to the deep ocean. But what if it didn't? Awkwardly, since 2004 the oceans are rising slower than previously, despite all that CO_2.[22] One study claimed the ocean has warmed by 240 sextillion joules since 1955, which sounds a lot more exciting than it is.[23] It translates to just 0.09°C in 50 years, and it's a rather brave and ambitious claim that we can measure the ocean temperatures to one hundredth of a degree even today, let alone in 1960. As far as measurements go, we left the Dark Ages of ocean heat with the 2003 ARGO program—a global array of more than 3,000 free-drifting profiling floats that measure the temperature and salinity of the upper 2000 metres of the ocean. Now there is almost one thermometer for each 200,000 cubic km of ocean. Is this what 95 per cent certainty looks like?

As it is, the ARGO data that has come in is like the thermometers on the surface, like sea-level measures, and like radiosondes in the upper atmosphere it doesn't show as much heat as the models predicted.

Instead of saying 'the pause' fits with 'missing energy' which fits with 'missing sea level rises' the excuses pile on excuses. What happened to the missing sea level rise? Apparently the ENSO effect dumped it on Australia.[24] So sea levels need adjustment too. When the data doesn't fit, we don't adjust the model, but we do adjust the data. What bad luck—all the major instruments are cold biased. What are the odds?

In any case, excuses for the pause don't solve the other flaws. The models not only fail on global scales, but on regional, local, short term,[25] polar,[26] and upper tropospheric scales too.[27] They fail on humidity, rainfall, drought and they fail on clouds.[28] The common theme is that models don't handle water well. A damn shame on a planet covered in water.

These doctors of dead science were surely given wings by a religious faith in their own insight. Only the true believers could believe thousands of instruments are biased against them (and in the same direction) and know that their 95 per cent certainty hides in the deep abyss. The science 'may be settled' indeed, but it settled somewhere in the Mariana Trench.

95 per cent certainty means 'discrepancies', 'surprises' and 'inconsistencies'

These same people, below, endorse the 95 per cent certainty. Here are their words on the differences between modelled and observed trends on the most influential feedback system in the climate models:

> Surprisingly, direct temperature observations from radiosonde and satellite data have often not shown this expected trend.[29]

> [T]he tropical troposphere had actually cooled slightly over the last 20 to 30 years (in sharp contrast to the computer model predictions) ...'[30]

> (Most) models overestimate the warming trend in the tropical troposphere ... The cause of this bias remains elusive.[31]

Shh, don't mention the water

To state the bleeding obvious, Earth is a water planet. Water dominates everything and it's infernally complicated. Water holds 90 per cent of all the energy on the surface,[32] and both NASA and the IPCC admit water vapour is the most important greenhouse gas there is.[33] They just don't seem inclined to produce posters telling us this is a humidity crisis, or that water *is pollution.*

The untold horror of humidity

Floating invisible water molecules are up to 100 times more abundant than CO_2 (literally 40,000ppm vs 400ppm). And water vapour absorbs and emits across wider bands of the infrared spectrum as well. Not to mention that there are even pools of liquid H_2O known to exist on the Earth's surface—like one called the Pacific. Meanwhile about 13,000 cubic kilometres of liquid and solid water is suspended in giant fluffy clumps that cover 60 per cent the Earth.[34] Each different incarnation of water has a different effect on the climate. Down on the surface, liquid water is dark and absorbs almost all the energy arriving. Solid water acts in exactly the opposite way—ice is like a mirror bouncing the energy back to space. Up in the air, thick white low clouds cool the planet by shading it, while thin high ice clouds have a net warming effect. There is no end to the contradictions. And you don't need to be a scientist to know that cloudy humid nights are warmer, while cloudy damp days are cooler. Dry air means temperatures swing more from hot to cold, while humid air keeps temperatures stable. The effects are so large none of us need a thermometer to know this.

When dada science became surreal science—the 'hot-spot' lives on

Even by 1990, the first searches for the hot-spot were hinting that it wasn't happening. For the next twenty years scientists re-analysed the weather balloons in dozens of papers to correct for every possible cooling bias they could find. With that path exhausted, things got more creative. In paper after paper the hot-spot kept being 'found' (albeit in odd conditional ways). Absent from the paper-flow was a single paper from those same scientists that headlined that it had gone missing. Pretty much the only time anyone admitted it was lost was in the introduction to a paper where they thought they had found it. Confirmation bias anyone?

Lessons in marketing zombie-science

It's been a dedicated, relentless quest to resuscitate the meme that died a thousand deaths. Those who were convinced in the theory really had nothing to work with (except a lot of money) but somehow they managed to keep the fear in play.

#1 Start with money

It takes a lot of money to keep a really silly idea afloat. (It's best if it's someone else's money and even better if you don't have to pay it back—thank the taxpayers of the Western world).

Sceptics are largely volunteers, and even the largest and most well recognised sceptical organisation—the Heartland Institute—runs on a small budget of around six or seven million dollars annually (for all its projects), yet the government gravy flows over believers like the Amazon river. Over $100 billion in scientific research buys a lot of irrelevant repetitive 'me-too' type of papers.[35] Each of those papers gets its own press release. To some extent, academia and science publishers are de facto advertising agencies, and they only have one customer—the government.

When the carrot is a $2 trillion global carbon market, it even brings out the green side of investment bankers. Deutsche Bank were so concerned about the environment they paid for a 70-foot-high carbon clock of doom in New York. (When will the bankers build a whale clock wall? When they can trade Humpback Credits.) This debate is so paranormal,

Deutsche bank didn't think the IPCC, UN, NOAA, NASA, and worldwide academia were doing enough to defeat sceptics and even issued a 50-page science report themselves called *Climate Change: Addressing the Major Skeptic Arguments*.[36] Is it a coincidence that in March 2009, Deutsche Bank had about $4 billion under management involving climate change?

Meanwhile the same teams of intrepid journalists who denounced and hunted Exxon for funding sceptics had no problem at all with bankers promoting believers. There were no headlines 'Bankers profit from Carbon Scare' or 'Deutsche protects market with Scare-Mongering Report'. Presumably journalists felt the banks were just interested in saving the planet.

Scientific American lauded a study showing as much as $558 million was funnelled to almost 100 'climate denial' organisations over seven years.[37] It was published in *Climatic Change*, because, after all, right wing think tanks are a recognised climate force.[38]

While Greenpeace was complaining about the Koch brothers controlling the climate debate with $67 million,[39] the renewables industry was quietly spending nearly a billion dollars a day.[40]

#2 Wordsmith—leave no definition intact

We think through our words, so clear logical thinking requires accurate English. But if your aim is marketing, not logic, accurate words are the enemy, and foggy text is your friend. Any word can be abused and reused. The practise is rife—indeed it starts and ends with abuses of language. The entire debate between scientists is reframed as a non-contest between 'experts' and 'climate deniers'. Don't ask anyone to define a climate denier, because literally it doesn't exist: no one denies we have a climate. Even John Cook, who wrote an entire book on the topic of 'deniers,' admits 'there is no such thing as climate change denial.'[41] Despite that, he doesn't seem to be in a hurry to fix his site, his papers, or his book.

It's as if a Wimbledon finalist declared they won before the game even started—because the other guy is a ball-denier. It wouldn't work in tennis, but in a science debate, the ambit claim fools professors and prime ministers alike. These spectators seemingly want to watch the contestants throw names at each other, instead of the ball. It's a parody in action.

Scientifically, things are so dismal that climate scientists are not even trying to kick a goal anymore. For them success now is when the ball can't be said to have missed. Yet.

#3 Sell the 'simplicity'—hide the unknowns.

Would you buy simple physics from this man? 'The science is settled,' said Al Gore.[42] It's 'simple physics' says Lord Rees, the President of the Royal Society.[43] Both of them salesmen.

The physics of CO_2 is simple, but the fine print on the models is that doubling CO_2 will only lead to 1.2°C of warming.[44] No catastrophe. You can ask James Hansen or the IPCC.[45] Did they forget to mention that all the disastrous predictions—two, four, six, or eleven-hyperbolic-degrees—come from assumptions about what humidity and cloud feedbacks will do? Repeat after me: physics points at one degree, everything above that is a vaporous damp guess.

While internet trolls make out that the sceptics deny basic laws of physics, a sea of climate scientists stand silently by keeping their error bars cloaked. Don't mention the feedbacks. Don't mention cloud microphysics either, and definitely don't mention *humidity*.

To be sure, greenhouse physics is fairly settled, but the climate system is a mess. Temperatures are difficult to predict, and nearly everything else is harder. The uncertainty monster practically eats rainfall projections for breakfast.

The climate billboards point at one small process, but the outcome depends on the whole system. Imagine a log cabin in the Antarctic

Circle. Closing the window doesn't make you much warmer when the front and back door are open and channelling the Katabatic wind. So is it with the Earth. Energy has other ways to escape.

It could be that we closed a window in a house with no walls. Time to panic?

#4 Fingerprint? What fingerprint?

When sceptics put the hot-spot predictions next to the radiosonde results and publicly asked where the signature effect of greenhouse gases was, pretty soon the response was to explain that it wasn't a fingerprint, because in theory any form of warming should cause a hot-spot. Given that it wasn't there, it rather suggested the theory might be wrong and water vapour wasn't amplifying anything much, and the climate was not that sensitive.

It also begged the question of why the largest body in US-climate-science used the word 'fingerprint' 74 times in their 2006 synthesis and assessment report.[46] This fingerprint they discussed promised to show unmistakably that anthropogenic forcings produced a different pattern to natural forces. The predictions were published in full glorious color in Chapter 1. Curiously, the contradictory results were also published in the same report but four chapters and 116 pages apart.[47] In 2006 the models showed that *only* greenhouse gas changes could cause a hot-spot. But there was no sign of the hot-spot. Not even remotely. Which raises the question of whether the same scientists would have still called it a fingerprint, had it turned up.

#5 Discover 'uncertainty' and rejoice!

A team of sceptical scientists published a paper showing that all four of the major global temperature datasets disagreed with the crucial modelled trends in the tropical troposphere.[48] The response of the modellers was to publish, with much hullabaloo, a refutation with seventeen authors. This was advertised as 'resolving a long-standing conundrum in climate science', but Santer et al. did not have new data (they didn't even use data after 1999).[49] Their great discovery was essentially to find uncertainties. By showing that we were even less sure of the results, we were therefore more sure the models were not proven failures. This was trumpeted as a success. They were stretching the error bars so wide that one data set of four measurements might overlap slightly with some model predictions.

Two years later a different team of sceptical scientists used the same techniques as Santer et al. with more recent data as well and got a very different result.[50] They concluded that the models were wrong (but only by 400 per cent).

#6 Why not use wind speeds to measure temperatures?

To estimate temperatures 10 kilometres over the tropics, in at least one paper keen researchers threw away the thermometer information and studied windshear instead.[51]

Temperature sensors had been specifically designed and calibrated to measure temperatures, but they weren't getting the right answer. Apparently things like radars, GPS-tracking and ways of measuring wind speeds are accidentally better at measuring temperatures than the thermometers. (If only we'd known. Think of all the money we wasted on all those thermometers …)

#7 When yellow is red

If there is one episode that really captures the state of the non-science, it's the point where the Emperor tells the world that yellow is truly red, and no one disagrees. I'm thinking of Sherwood et al.'s 2008 paper: where you can find a hot looking-spot in a space with a zero degree trend.[52] One graph looked for all the world like the predicted hot-spot graph and it was used that way in the blog-world. But the colors in the scale were shifted so that even 'no warming' would be marked with a hot orange-red. The color scale was not just counter intuitive; it actively prevented anyone from comparing the trend in the upper troposphere with the surface. Any warming trend at all was blurred into a similar shade of red. Somehow this was published in a peer reviewed journal.

John Cook, who now works for the University of Queensland, used the Sherwood graph in a publication to try to refute the *Skeptics Handbook*.[53] When I pointed out the color scale trick, Cook didn't say a thing publicly, didn't protest, and dropped it from an expanded, similar booklet that came out six months later. He continued to defend other work by Sherwood. That's how it goes. Perhaps it was just a printing mistake?

Or perhaps not. The graph has the words 'hot spot' added over one point which is the also the wrong point (it's too low in the atmosphere). And Professor Sherwood is listed as an advisor on the guide.

#8 Black belt graphs

The clear descriptive graphs of the hot-spot and the radiosondes were published in 2006 and 2007, but after the bad publicity, by 2013 the IPCC 'redesigned' them. Instead of an easy-to-read visual graph they split the atmosphere into four zones, included a lot of the unnecessary stratosphere, removed the altitude in kilometres from the right hand side, and generally complicated and reduced the discrepancy.[54] Instead of a rainbow of colors the Hadley radiosondes became a thin black line surrounded by colourful spaghetti. If this were an art movement it would be called 'Clutterist'.

Like tricks used to market things you don't need at Walmart, this publication achieved through graphic design what it failed to do with data. Another paper dismissed the radiosonde data as 'spurious'. The key graph in Dessler's 2010 study was one where all the radiosonde results get packed into one thin lonely line far from the model forecast, while the satellite data gets reanalysed—and displayed spaghetti style weaving together.[55] It creates the illusion that the 28 million independent radio-sondes are just an outlier.

#9 Pretend the hot-spot doesn't matter

It's the end-game stage. Now that almost all options are exhausted the latest and probably last tactic—plan Z—is to declare the hot-spot didn't matter after all, was never important and 'has no implication. Nil'.[56] That's the same Sherwood who said the water vapour feedback doubles the effect of any other warming. Why? He goes on to explain: 'Anyone who wants to argue that the "missing hot-spot" implies something as to the future (say, that global warming will be less than current models predict) needs to come up with an alternative model of climate.' In other words, our models don't work, but you need to make ones that do before you can criticise them.

We're setting national policy with unverified models that don't match the data. The answer is not to 'keep spending'—but to get the models to work.

#10 Call your opponents crazy conspiracy thinkers

When all else fails scientifically, it's time to use smears slurs, ostracism, and general character assassination. (Actually this was the first choice and used all along. It has just reached new heights of absurdity).

Essentially those who believe in the carbon-disaster are 95 per cent certain sceptics are big tobacco-funded anti-Semitic deniers who are so stupid they doubt the moon landing was real. One team of psychologists used thousands of taxpayer dollars to conduct an online survey.[57] The survey was aimed at sceptics but was posted almost entirely on sites that were virulently anti-sceptic. It's like trying to discover what Jews think, but only interviewing people in Gaza.

In the end, they got only ten anonymous responses from people who said they believed the moon-landing was faked, and only four of those also claimed to be sceptics. Based on this nano-fragment of reality, the team immediately issued press releases declaring sceptics were more likely to believe the moon landing was faked. Careful investigative journalists swallowed the story whole and it was published in the great masthead *The Guardian*.

It took the researchers another seven or eight months to actually check, correct and review their work and get it published. (Lucky no one believes anything they read in the mainstream media, isn't it?) Naturally, scientists everywhere protested at the statistical incompetence and inept design. In response, the same team of psychologists diagnosed those making the complaints as suffering from various forms of 'conspiracy ideation'. They published their online diagnoses in a second paper, 'Recursive Fury,' which was so hopelessly ethically and scientifically compromised that it was removed from the journal's site within weeks, and officially retracted completely a year later.[58]

Was the point of this research to advance human knowledge or to advance a cause?

Carbon dioxide is not a dangerous pollutant, not unless you measure ground temperatures in car parks, and tropospheric temperatures with wind gauges. In the absence of better information, based on what we have, the simplest explanation is that man-made greenhouse gases have minor warming effects.

12 Forecasting global climate change

Kesten C. Green & J. Scott Armstrong

Warming by 2070, compared to 1980 to 1999, is projected to be … 2.2 to 5.0°C.

—CSIRO[1]

By 2100, the average U.S. temperature is projected to increase by about 4°F to 11°F.

—US Environmental Protection Agency[2]

If we do not cut emissions, we face even more devastating consequences, as unchecked they could raise global average temperature to 4°C or more above pre-industrial levels by the end of the century. The shift to such a world could cause mass migrations of hundreds of millions of people away from the worst-affected areas. That would lead to conflict and war.

—Nicholas Stern, Baron Stern of Brentford[3]

Forecasts such as these are made by scientists and repeated by the political leaders they advise.[4] The principal source of the forecasts is the United Nation's Intergovernmental Panel on Climate Change (the IPCC). The IPCC's forecasts are the product of a collaboration of scientists and computer modellers working for lobbyists, bureaucrats, and politicians.[5] The forecasts of dangerous manmade global warming and its consequences are made with great confidence, as are recommendations of actions to counter the forecasted danger.

History is replete with experts making confident forecasts. The record also shows that the accuracy of such forecasts has been poor. Consider, for example, Professor Kenneth Watt's forecast of a new Ice Age in his 1970 Earth Day speech at Swarthmore College:

> The world has been chilling sharply for about twenty years. If present trends continue, the world will be about four degrees colder for the global mean temperature in 1990, but eleven degrees colder in the year 2000. This is about twice what it would take to put us into an ice age.

Watt is not unusual among experts in making confident forecasts that turn out to be wrong. Evidence from research on forecasting shows that an expert's confidence in making forecasts about complex uncertain situations is unrelated to the accuracy of the forecast.[6] Those who believe that we can learn to avoid poor forecasts from history may wish to consult the diverse examples in Cerf and Navasky's 1984 book *The Experts Speak.*[7]

We suggest that government policy makers and business managers consider whether the IPCC's forecasting methods are valid before they consider making decisions on the basis of the forecasts. To that end, we examine whether or not the IPCC's forecasts of dangerous manmade global warming are the product of scientific methods.

We then investigate whether alternative hypotheses of climate change provide more accurate forecasts than the dangerous manmade global warming hypothesis. Specifically, we test forecasts from the hypothesis of global cooling and from the hypothesis of climate persistence. We then make forecasts of global average temperatures for the remaining years of the twenty-first century and beyond using an evidence-based forecasting method.

Finally, we ask whether the IPCC forecast of dangerous manmade global warming is a new phenomenon. To answer this question, we use the method of structured analogies to seek out and analyse similar situations.

Are the alarming forecasts the product of scientific forecasting methods?

The IPCC forecasts are derived from the judgments of the scientists that the IPCC engages. Computer modellers write code to represent the scientists' judgments that, in turn, provides long-term forecasts of global mean temperatures. Is this use of expert judgment a valid approach to climate forecasting?

For nearly a century, researchers have been studying how best to make accurate and useful forecasts. Knowledge on forecasting has accumulated by testing multiple reasonable hypotheses about which method will provide the best forecasts in given conditions. This scientific approach contrasts with the folklore that experts in a domain will be able to make good forecasts about complex uncertain situations using their unaided judgement, or using unvalidated forecasting methods.[8]

Scientific forecasting knowledge has been summarised in the form of principles by 40 leading forecasting researchers and 123 expert reviewers. The principles summarise the evidence on forecasting from 545 studies that in turn drew on many prior studies. Some of the forecasting principles, such as 'provide full disclosure' and 'avoid biased data sources,' are common to all scientific fields. The principles are readily available in the *Principles of Forecasting* handbook.[9]

We used that knowledge to assess whether the procedures described in the 'Climate Models and their Evaluation' chapter of 2007 IPCC Assessment report amounted to scientific forecasting.[10] To do so, we first examined that IPCC chapter's references to determine whether the authors had relied on validated forecasting procedures. We found no references to validation. We then sent emails to all of the authors of that section for whom we were able to obtain email addresses,[11] asking for references for credible forecasts of global average temperatures and the methods used to derive them. The few useful responses we received referred us to the 'Climate Models and Their Evaluation' chapter or to works that were cited in it.

We then audited the IPCC forecasting procedures using the Forecasting Audit Software available on ForPrin.com. Our audit found that the IPCC followed only seventeen of the 89 relevant principles that we were able to code using the information provided in the 74-page IPCC chapter. Thus, the IPCC forecasting procedures violated 81 per cent of relevant forecasting principles.[12]

It is hard to think of an occupation for which it would be acceptable for practitioners to violate evidence-based procedures to this extent. Consider what would happen if an engineer or medical practitioner, for example, failed to properly follow even a single evidence-based procedure.

We analysed the IPCC's forecasting procedures to assess whether they followed the Golden Rule of Forecasting. The Golden Rule of Forecasting requires that forecasters be *conservative*. This means that they should use procedures that are consistent with knowledge about the situation and about forecasting methods. The Golden Rule is the antithesis of the common anti-scientific attitude that 'this situation is different,' which leads forecasters to ignore cumulative knowledge.

The Golden Rule is a unifying theory of how best to forecast. The theory has been tested for consistency with the evidence in a review of the literature from all areas of forecasting that found 150 studies relevant to the Golden Rule. The studies provided findings from experiments on the effect of conservative procedures compared to unconservative ones on forecast accuracy. All of the evidence was consistent with the Golden Rule.

To assist forecasters, the evidence on the Golden Rule is summarised in the form of 28 guidelines, including 'avoid bias by specifying multiple hypotheses and methods' and 'select evidence-based methods validated for the situation'.[13] The median reduction in forecast error from following a Golden Rule guideline, rather than common practice, is 25 per cent. That is, error was reduced by one quarter.

CLIMATE CHANGE: THE FACTS

We found that the IPCC procedures violated all nineteen of the Golden Rule guidelines that are relevant to long-term climate forecasting, including 'be conservative when forecasting trends if the series is variable or unstable' and 'be conservative when forecasting trends if the short and long-term trend directions are inconsistent.' As a consequence of the Golden Rule violations, the IPCC forecasts are a product of biased forecasting methods.

Are forecasts of dangerous global warming nevertheless valid?

Having established that the IPCC *forecasting procedures* are invalidated and are inconsistent with scientific forecasting knowledge, we investigated whether it would be possible to test the validity of the *forecasts*.

The most recent global warming scare started around 1976, so testing the validity of short-term forecasts against the few years since then is possible. Such a test is limited, however, given that it is not unusual for temperatures to trend up or down, on average, for several years. Also, policy makers and investors who consider large expenditures that are costly to reverse are concerned with long-term trends. We therefore devised tests of the validity of the IPCC model's short- and long-term forecasts that made extensive use of available data.

In 1999, Michaels explained that short-term events were responsible for recent elevated temperatures and offered an early test of the IPCC's short-term forecasts in the form of a bet that temperatures would go down in the next ten years.[14] No one took the bet ... and temperatures went down.

Over the past nearly two decades, atmospheric CO_2 concentrations have risen while global temperatures have remained flat. Despite the disconfirming evidence, the IPCC claims to have become even more confident about the man-made global warming hypothesis and they continue to forecast dangerous warming. The IPCC's response is typical of how people tend to react when their forecasts are wrong: by having an even stronger belief that they will be proven correct.[15] Moreover, scientists, like other human beings, tend to reject evidence that contradicts their beliefs.[16]

By 2007, there still had been no proper validation of the IPCC's forecasts. To generate interest in the importance of validation, one of us (Armstrong) proposed a bet to former U.S. Vice President Al Gore that a 'no-change' forecast of global average temperature would be more accurate than any model or forecast that Mr. Gore would support. Gore, advised by Professor James Hansen,[17] was at the time warning that a 'tipping point' in global temperatures was imminent. In contrast to Gore's expectation of supporting evidence soon, Armstrong expected that a much longer period would be needed to obtain a clear result due to natural variations. Armstrong nevertheless proposed a ten-year bet on the assumption that a shorter term would generate more interest, despite estimating that he had a one third chance of losing.

In order to have an objective standard against which to compare forecasts from the alternative hypotheses, the bet uses the University of Alabama at Huntsville (UAH) lower troposphere series.[18] As of May, 2014, the errors from the IPCC's business-as-usual forecast of +0.03°C per year—standing in for Mr. Gore's tipping point due to his unwillingness to take the bet—were more than 27 per cent larger than the errors from Armstrong's bet on the no-change forecasts.

The models that the IPCC uses for forecasting are based on the beliefs of some scientists that exponentially increasing levels of CO_2 in the atmosphere will cause global mean temperature to increase at a rate of at least 0.03°C per year. That figure has been the central forecast of the IPCC since 1990.[19] Because CO_2 levels have been increasing exponentially since the beginning of the Industrial Revolution, the IPCC model would seem to apply over this whole period.

We tested the validity of the IPCC model for forecasting horizons up to 100 years using the data on global mean temperatures that the

IPCC use: the U.K. Met Office Hadley Centre's HadCRUT3 series.[20] The Hadley temperature series are derived from selected weather stations and sea surface records that are adjusted and aggregated to provide proxy average global temperatures. We derived rolling IPCC-model forecasts of the HadCRUT3 series starting from the year 1851, and ending in the year 1975, before the most recent global warming trend commenced. The forecasting procedure was simple, and is consistent with the IPCC's published business-as-usual forecasts: we added 0.03°C to the previous year's actual temperature to derive a one-year-ahead forecast, and then added the same figure to the forecast for the previous year for each subsequent forecast horizon out to 100 years. By repeating this procedure for each subsequent year, we obtained 125 one-year-ahead forecasts, 124 two-year-ahead forecasts, and so on, up to and including 26 forecasts for 100 years ahead.[21]

Given that the HadCRUT3 temperature series trends broadly upwards,[22] one would expect the IPCC-model forecasts that we generated to track the HadCRUT3 series quite well. To determine whether the dangerous global warming hypothesis is a credible one, however, it is necessary to test the forecasts against forecasts from alternative hypotheses, and to do so using scientific forecasting methods.

In the 1960s and early 1970s, scientists warned of a new ice age.[23] The scientists provided hypotheses to support their belief that this time the climate really had changed. Some scientists still advance the cooling hypothesis.[24]

Yet despite these forecasts of cooling, starting in the mid-to-late 1970s there was actually a warming trend, and warming alarmists began to inform us that virtually all scientists now subscribed to the dangerous man-made global warming hypothesis. The claim of near unanimity of scientific opinion has been discredited by Legates et al.,[25] however, and stands in contrast to the 31,487 U.S. scientists who have publicly signed a statement that they consider the dangerous manmade global warming hypothesis inconsistent with the evidence.[26]

While scientists who predict warming and those who predict cooling provide reasons for their hypotheses, their reasons have been indecisive. In any event, science does not advance by asking scientists to vote on hypotheses, but by testing them in competition with alternative reasonable hypotheses.[27]

We tested a cooling hypothesis of 1°C cooling per century against the HadCRUT3 global temperature data. The forecast of cooling is consistent with the various alarms over impending new ice ages that have occurred over the last 100 years and longer, including those mentioned above.[28] And the rate is arguably consistent with the understanding of scientists who consider that the Earth is still experiencing a cooling period, albeit with fluctuations, that commenced around 4,000 years ago.[29]

For horizons from one to 100 years from the year 1851 to the year 1975—7,550 forecasts in total—the average absolute errors of the 0.03°C per year warming forecasts and of the 0.01°C per year cooling forecasts increase as the forecast horizon increases (see Figure 1). Because our tests use historical data known to exhibit a warming trend, the warming model has an unfair advantage in this test. Despite that advantage, across all forecast horizons, the average errors of the warming forecasts are more than twice as large as the errors from the relatively more conservative cooling hypothesis. Remarkably, the natural cooling forecasts are more accurate than the dangerous warming forecasts for all forecast horizons.

The global warming and cooling hypotheses were developed without the aid of scientific forecasting. To develop a credible forecasting method against which to benchmark the warming and cooling hypotheses, we needed a model that was both consistent with evidence-based forecasting principles and with evidence on climate change. With that in mind, we asked climate expert and astrophysicist Soon to collaborate with us to develop a model and validation tests.[30]

With Soon, we established that the state of knowledge about the causes of climate change was such that it would be inappropriate to develop a causal model. The strength and even direction of proposed causal relationships, including with CO_2, are much disputed among leading climate scientists.[31] For example, in 1972 Kukla and Matthews reported from a meeting of climate scientists that :

> one conclusion reached at the session was that there is no qualitative difference between the climatic fluctuations in the 20th Century and the climatic oscillations that occurred before the industrial era. The present climatic trends appear to have entirely natural causes, and no firm evidence supports the opposite view.[32]

A more recent analysis of two 3,000-year temperature proxy series comes to the same conclusion.[33]

We concluded from forecasting principles that because knowledge about climate change is so poor, forecasts from a no-change forecasting model would be more accurate than forecasts from methods that attempt to incorporate knowledge that is tentative at best. Depending on the situation, the appropriate no-change model might be one that forecasts that the level (e.g. current temperature) will not change, that the trend will not change, or even that the rate of change will not change. For forecasting long-term global temperatures, we determined that the benchmark model that is most consistent with the state of knowledge is one that forecasts no change in the level; in other words, no trend.

We compared the forecasts from the no-trend model with the forecasts from the cooling and warming hypotheses. We found that the average error of the no-trend forecasts was smaller than the average errors of both the warming and the cooling forecasts for *all* forecast horizons (Figure 1).

The average errors of the warming forecasts (dashed line) and the cooling forecasts (dotted line) over the short-term (one to ten years) were

Figure 1: Average absolute errors of 0.03ºC warming, 0.01ºC cooling, and persistence forecasts

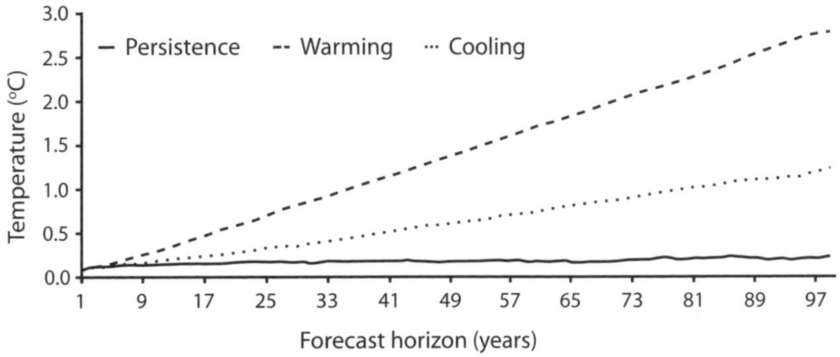

Forecasts for 1851 to 1975 by forecast horizon.   Source: K.C. Green and J.S. Armstrong

45 per cent and 10 per cent larger, respectively, than the average errors of the no-trend forecasts (solid line). The average error of the no-trend forecasts for the longer-term horizons, from eleven to 100 years, was roughly one-quarter of the average cooling forecast error, and one-eighth of the warming forecast error. In absolute terms, the average errors of the no-trend forecasts were less than 0.20°C for all horizons out to 75 years; beyond that, the average errors did not exceed 0.24°C . The small and steady forecast errors from the persistence model suggest that the Earth's climate is remarkably stable over human-relevant timescales. This is particularly remarkable given the claims by warming alarmists that we have been experiencing 'unprecedented' changes in the climate over the period of the test.[34]

Very long-term testing of predictive validity

In order to assess the validity of the hypotheses over very long horizons, we tested the accuracy of forecasts from warming, cooling, and no-trend model hypotheses against the Loehle series of proxy annual temperatures.[35] Proxy temperature data are obtained from naturally occurring records of biological and physical processes that vary with temperature.

The Loehle series was constructed from eighteen series obtained and calibrated by other researchers who used such proxy records as boreholes and pollen counts that each covered most of the Common Era and, between them, covered much of the globe. The resulting Loehle series extends from AD 16 to AD 1935, allowing us to test forecasts from variations of the hypotheses for horizons of up to nearly 2000 years. The series includes the Medieval Warm Period and the Little Ice Age. Evidence suggests that the current climate is not as warm as that of the Medieval Warm Period when cows grazed and willows grew in Greenland and seals basked on the shores of Antarctica.[36]

A forecaster living 100 years after the beginning of the Loehle series in AD 115 might reasonably have forecast that the average temperature trend that had prevailed over the previous 100 years, an increasing one of roughly 0.003°C per year (0.3°C per century), would prevail indefinitely. Indeed, some researchers have suggested that the Earth has been warmed by human activity for at least 5,000 years.[37] The errors of the warming forecasts increased as the forecast horizon lengthened as the dashed line in Figure 2 shows.

Figure 2: Absolute errors of warming, cooling, and no-trend forecasts

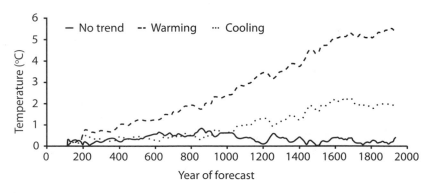

By year from AD 115 to AD 1935, in degrees Celsius.   Source: K.C. Green and J.S. Armstrong

180

A competing forecaster in AD 115 might well have reasoned from the knowledge of the time that our sun is like a large fire that must slowly burn down. Given that temperatures over the previous century had been trending more upwards than downwards, she might have proposed that while the Sun's fire may splutter and flare up from time to time, there would be a long slow decline in the energy emitted. With these observations in mind, she might have forecast that the average temperatures would trend downwards at the relatively more conservative rate of 1°C per millennium or 0.001°C per year on average—a much more conservative forecast than those of the first millennium warmer and of the twentieth century warmers and coolers described above. While the errors of her cooling forecasts increased only slowly out to the year AD 750, beyond that year the errors of her forecasts tended to increase as the forecast horizon lengthened (dotted line in Figure 2).

We compared the records of the warming and cooling hypotheses forecasters with the record of our benchmark no-trend hypothesis in the form of a forecaster who predicted that the global average temperature for the 1,820 years from AD 116 to 1935 would be the same as the AD 115 average. The solid line in Figure 2 shows the errors of the no-trend forecasts by year, one forecast error per year. The modest size of the errors and the lack of even a very small persistent trend in them suggest that there have been no changes in the climate system. In other words, the claim that 'things are different now,' although often made in relation to forecasting in many fields, is once again unsupported. Over longer policy-relevant periods, annual global mean temperatures are highly stable.

Even with a much more conservative forecast warming rate (one-tenth that of our previous tests), when applied to this series the warming hypothesis again performed relatively poorly. The average error of the 1,820 years 0.3°C -per-century warming forecasts was more than nine times the average error of the no-trend forecasts. Again, the errors increased with the forecast horizon. For example, the errors of warming forecasts for the fourth century made in AD 115 were nearly three times larger than the errors of the no-trend forecasts. The equivalent figures for the eighth, twelfth, sixteenth, and eighteenth centuries were four, fourteen 23, and 27 times larger. The findings are consistent with those of Green, Armstrong, and Soon.[38]

These findings from a long period of varied climate, then, are consistent with those of our analysis for the 1851 to 1975 warming period above: the more conservative hypothesis and forecasting method provides the more accurate forecasts. In particular, the most conservative model, the no-trend model, has greater predictive validity than long-term trend models under diverse conditions. No matter when one starts forecasting and no matter how global average temperature is estimated, the evidence-based persistence model produces by far the most accurate forecasts. The findings on the accuracy of forecasts from long and short-term tests of the alternative climate change hypotheses are summarised in Table 1 in the form of Relative Absolute Errors (RAEs). The reported RAEs are the absolute error of the forecasts from the hypothesis relative to the corresponding absolute error of the forecasts from the persistence (no-change) model over the forecast horizon. Thus, a figure of 0.5 means the error was only half as big as that from the persistence model forecast, and 2.0 means it was twice as big.

Evidence-based climate forecasts for the 21ˢᵗ century

Our testing used alternative data sources, different time periods, different starting points, and different horizons. The findings were always the same. Forecasts from the more conservative cooling hypotheses were more accurate than forecasts from the warming hypotheses. Forecasts from the most conservative hypothesis, the no-trend model, were always much more accurate. The no-trend model is consistent with evidence-based forecasting principles and with the state of knowledge about the behaviour of the Earth's climate.

Table 1: Relative accuracy of forecasts from alternative climate change hypotheses, warming, and cooling, versus persistence

Data Series (Test period)	Forecast Horizons (years)	Forecasts (number)	Hypothesis (Rate, °C p.a.)	Relative Absolute Error (v. Persistence=1)
UAH (2008-2014)	1-6$^{1/3}$	76*	Warming (0.03)	**1.27**
			Cooling (0.01)	1.05
HadCRUT3 (1851-1975)	1-10	1,205**	Warming (0.03)	**1.45**
			Cooling (0.01)	1.10
	11-100	6,345**	Warming (0.03)	**8.14**
			Cooling (0.01)	3.62
Loehle (116-1935)	1-1,820	1,820	Warming (0.03)	**9.01**
			Cooling (0.01)	3.16

*Monthly forecasts. **Successive updating used.*

The IPCC's alarming-warming model is not. Consistent with knowledge about the proper model for this situation, the predictive validity tests finds no support for the global warming hypothesis for forecasting global mean temperatures over this century and beyond.

Our forecasts for each year's global average temperature for the 100 years to 2113 are that they will be the same, more or less, as the 2013 global average temperature. We suggest that our forecasts should be monitored against the University of Alabama at Huntsville's (UAH) lower troposphere temperature series because this satellite-based measure provides a better assessment of the global average than the Hadley (HadCRUT3) series, and because it is fully and openly documented and is, therefore, less likely to be biased.

Perhaps it is possible to improve on the already very accurate long-term temperature forecasts from the no-trend model, for example by estimating the global average temperature level from a weighted average of temperatures over recent years, rather than from only the latest year. We have not attempted to improve upon our very simple no-change model, however, because the errors of the forecast from the model are too small to be of concern to policy makers and business decision makers; the no-trend model forecasts are more than good enough.

What do previous environmental alarms tell us?

Having investigated the forecasting procedures behind the IPCC forecasts and found them to lack validity, and having found the forecasts to be much less accurate than no-change forecasts, we were concerned that governments are taking the dangerous man-made global warming alarm seriously, to the extent that they have already implemented costly policies and regulations. We decided, therefore, to examine whether the global warming alarm is an unusual phenomenon. To do so, we used the structured analogies approach.[39]

The structured analogies method involves asking experts to think of similar situations to the situation of interest. Information is then obtained about the outcome of each analogy. We had previously tested the structured analogies method for forecasting complex situations involving interactions between parties with conflicting interests, including a special interest group occupying a public building and demanding taxpayer funding, and an international crisis over access to water. The research found that a structured search for and analysis of analogous situations produces forecasts that are much more accurate than the usual method of asking experts what they think will happen.[40] Other researchers have subsequently found the method useful for forecasting the outcomes of policy initiatives.[41]

With the help of domain experts we have, to date, identified 26 analogous situations.[42] They all began with an allegedly portentous incident or with claims that an apparent trend was ominous. Searches for evidence supporting each alarm followed, along with calls for government action. In no case was there recourse to scientific forecasting. The fact that we were able to identify as many environmental alarm analogies as we did, and the frequency with which they have occurred in recent times, suggest that they are a common social phenomenon and that the global warming alarm is not at all unusual. More generally, it is another example in a long history of calamity forecasts similar to those described in MacKay's 1841 book, *Extraordinary Popular Delusions And The Madness Of Crowds.*[43]

Evidence on the nature and outcomes of all 26 analogies is provided in our online working paper at publicpolicyforecasting.com.[44] We welcome further evidence on each of these analogies, invite others to submit their ratings of the analogies for publication at publicpolicyforecasting. com, and encourage others to propose other environmental alarms in case we have missed important analogies.

What were the outcomes of the alarms? The forecasts of harmful outcomes all turned out to be wrong. For the 23 alarms that resulted in government actions, the measures that were taken caused harm in twenty cases. The alarms faded from public attention slowly over time, but harmful policies have remained in many cases. We suggest using the Golden Rule of Forecasting to identify and to expose such false alarms, and to thereby help to minimise the harm that they cause.

Conclusions

Climate has varied in the past and can be expected to do so in the future. Mankind has adapted to both cool and warm periods, and trade and economic growth over the past 300 years has greatly increased our ability to do so. In that context, forecasts of climate are of little value unless they are for a strong and persistent trend, *and* are accurate.

The IPCC 'forecasts' are for a strong and persistent trend, but they have been inaccurate in the short term. Moreover, there is no reason to expect them to be accurate in the longer term. The IPCC's forecasting procedures violate all of the relevant Golden Rule of Forecasting guidelines. In particular, their procedures are biased to advocate for the hypothesis of dangerous manmade global warming.

We found that there are no scientific forecasts that support the hypothesis that manmade global warming will occur. Instead, the best forecasts of temperatures on Earth for the twenty-first century and beyond are derived from the hypothesis of persistence. Specifically, we forecast that global average temperatures will trend neither up nor down, but will remain within 0.5 °C (1°F) of the 2013 average.

This chapter provides good news. There is neither need to worry about climate change, nor reason to take action.

The
climate change
movement

13   The search for a global climate treaty

Rupert Darwall

'You sealed a deal,' UN Secretary General Ban Ki-moon told exhausted delegates at the end of the December 2009 Copenhagen climate conference.[1] Only they hadn't. There wasn't a treaty or even a draft text resembling one. Instead there was a toothless accord which merely listed countries' voluntary pledges. 'We will try to have a legally binding treaty as soon as possible in 2010', Ban promised reporters.[2] There wasn't one in 2010, 2011, or in 2012, when the Kyoto Protocol's first commitment period ran out. The destination of the Bali Road Map, adopted in 2007, had been to arrive at agreement on a son of Kyoto before it expired. Now Bali looked like a road going nowhere.

Copenhagen was 'an incredible disaster', newly installed president of the European Council Herman van Rompuy told an American diplomat four days later, predicting that the 2010 climate conference in Cancún would be too. The Europeans had been 'totally excluded' and 'mistreated'. It was lucky he had decided to stay away. 'Had I been there my presidency would have been over before it began', van Rompuy confided. His top aide likened the prospect of the Cancún talks to a repeat of a bad film: 'Who wants to see that horror movie again?'[3]

As a means of cutting greenhouse gas emissions, the architecture of the climate change treaties had a fundamental defect. Ever since adoption of the Berlin Mandate, brokered by Angela Merkel in 1995, there had been a rigid bifurcation between industrialised and developing nations. Yet as everyone knew, it would be impossible to cut global emissions if China, India and other large emerging economies were excluded. At Bali, the Bush administration had insisted on a process that included the large developing economies. One way or another, the climate change Berlin Wall had to be torn down—not least to avoid repeating the fate of the Kyoto Protocol, as it remained highly improbable that the US Senate would ratify any treaty that did not include China.

The Obama administration accepted the Bush strategy and carried it forward. Copenhagen was the moment of truth which would test the proposition that China and India were willing to be legally bound to restrict their greenhouse gas emissions and, therefore, that a global treaty was possible. There was no ambiguity about the required outcome: China and India had to be in. While the US and Europe had a shared objective to find some way around the climate change Berlin Wall, the conclusions they drew were diametric opposites. This fundamental disagreement was to have profoundly adverse consequences for the chances of achieving any meaningful outcome from the post-Copenhagen negotiations.

Although the Obama administration had proposed a treaty with legally binding commitments in the run-up to Copenhagen, policy post-Copenhagen should be based on the reality of what had happened at Copenhagen. The US analysis, outlined by Obama's climate envoy Todd Stern in October 2010, took as its starting point that Kyoto's Berlin Wall paradigm was, in Stern's words, 'unworkable as a matter of both substance and politics:'[4]

[T]he United States, as well as a number of other countries, would not accept legally binding commitments unless China and other emerging markets did so as well, and they have made abundantly clear that they will not.[5]

Stern challenged the long-standing 'article of faith' that the world needed a legal treaty to govern international action. Instead of a legally binding treaty, Stern argued that negotiators should aim for an agreement that was politically and morally binding on the parties. Stern could also have said, but didn't, that such an international agreement falling short of a treaty would have the additional advantage of not needing a two-thirds vote in the Senate to be ratified. Indeed, avoidance of Congress has become a hallmark of the Obama administration's climate change policies, demonstrated by its recourse to the Environmental Protection Agency (EPA) regulation rather than cap and trade legislation.

Adoption of the voluntary accord at Copenhagen had been vetoed by Cuba and its South American allies, so the conference could only 'take note' of it. The immediate aim of US policy post-Copenhagen was to remedy this and get the accord formally incorporated into the climate change negotiating texts. In large measure, the US succeeded in achieving this at Cancún. Briefing the press in the run-up to the next conference, Stern called Cancún 'the most significant agreement since the Kyoto Protocol.'[6] According to Stern,

> We reached an agreement, which although it is not legally binding, it is a COP [Conference of the Parties] decision under a legally binding treaty, which is very serious and which covers more than 80 percent of global emissions as compared to a Kyoto agreement, which people are hoping will cover something in the order of 15 per cent.[7]

Stern's problem was that the rest of the world barely noticed. Such small-bore results were hardly commensurate with the soaring expectations created by Barack Obama's 2008 promise to slow the rise of the oceans and start to heal the planet. It was a strategy that came to grief at the 2011 Durban Conference of Parties (COP).

By contrast, the EU's goal was to re-litigate Copenhagen and find the Holy Grail of a binding treaty embracing all major emitters. At

Copenhagen, the West had tried to isolate China, offering $100 billion a year of climate finance to African and other least developed countries if China capitulated. Whereas the US realised that backing China into a corner hadn't worked then and wouldn't work in the future, the EU set about peeling off the coalitions that had veiled the full extent of Chinese intransigence.

It was the opposite of US climate diplomacy, which quickly grasped the requirement to develop an understanding with China on the parameters of any eventual deal. With the US and China jointly accounting for 44 per cent of global carbon dioxide emissions, developing bilateral agreement between the two nations was a *sine qua non* for a global deal. The US side worked intensively to create a strategic climate change dialogue with China. In a May 2010 op-ed in China's *Global Times* ahead of a Beijing session of the China Strategic and Economic Dialogue, Hillary Clinton wrote that the US and China had a 'unique responsibility' to lead efforts on climate change.[8] However, in the UN climate negotiating fora, the attempt would be bedevilled by mistrust and suspicion.

The EU on the other hand wanted to use the arena of the UN climate conferences to ratchet forward progress toward its goal of having a legally binding treaty ready for the Paris COP at the end of 2015. This involved cajoling, bribing—with promises of climate funding—and pressuring developing countries toward a destination they didn't want to reach. As Yvo de Boer, the recently departed UN climate convention executive secretary, explained in a frank interview just before Cancún, many developing countries were not deeply committed to the concept of green growth that climate change demanded:

> Although many nations pay lip service to this model, most of them, deep in their hearts are unsure. In fact many developing nations fear that the intent of the west is to use climate as an excuse to keep [them] poor and maintain the current status quo.[9]

In executing its strategy, the EU was suckered by habituation to its internal institutional processes—the intractable but irreversible process of reaching agreement within the 27-member bloc—which it projected onto the climate change negotiations, where positions can be reversed, apparent progress can be undone and what happens to be agreed at a particular moment does not determine the final outcome. The EU also held what it viewed as its trump card—unless other countries moved in its direction, it would not enter into a second commitment under the Kyoto Protocol.

This was presented as the EU doing the world a big favour. In reality, the EU was desperate for a rationale for persevering with Kyoto. In 2007, it had already committed itself to cut emissions by 20 per cent on 1990 levels by 2020. Pre-committing itself looked even more foolish as other developed countries confirmed they were not going to sign up to a second commitment period. Japan, Russia and Canada (which subsequently withdrew from Kyoto altogether) was followed by New Zealand, leaving Australia and Kazakhstan as the only non-European states to participate in the second commitment period.

At Cancún, the EU strategy met with some success. The pattern of the annual climate change conferences is that following a particularly disastrous COP, the next one is more harmonious, with participants sharing an interest in demonstrating to the outside world the viability and relevance of the talks. Unlike Copenhagen, there was an air of harmony at the closing plenary. Speaking for the Africa Group, Algeria said the COP had restored confidence in the process, observing that 'tonight God has been very close to Mexico.'[10] Earlier in the COP, many parties expressed support for a 'legally binding outcome' and the EU reiterated its willingness to sign up to a second commitment period in the context of a global, legally binding outcome.[11]

Apparent movement by India, China's core ally at Copenhagen, toward the EU's position created some excitement at Cancún and was severely criticised back home. Jairam Ramesh, India's environment minister, explained: 'We have to accept the changing global reality. G-77 [bloc of developing countries] has been vocally calling for [a] legally binding instrument.' At the same time Ramesh denied that this shift was going against India's fundamental position.

> The cabinet has given me a mandate not to agree any legally binding agreement at this stage … What I said was to keep India's options open in global climate negotiations. I don't want to put India in a corner. We should have room for manoeuvre for the next two years … I admit our position has evolved, keeping in mind India's changing global role.[12]

However Indian premier Manmohan Singh asked critics 'not to read too much into the statement.'[13] Green growth must have been far from Singh's mind. A month earlier, Singh had told India's energy firms to scour the globe for fossil fuel supplies as India's demand for them was set to increase by 40 per cent. 'Hydrocarbons will continue to be our major source of energy for quite some time', Singh said.[14] India had a different negotiator at the 2011 COP in Durban.

The Europeans arrived in Durban brandishing their big bazooka. Melodrama was the order of the day. The chief G-77 negotiator suggested Durban could be the place where the Kyoto Protocol dies—or 'be put on a respirator, just so it will not die here, so it will die somewhere else.' Host president, South Africa's Jacob Zuma and one of the BASIC four alongside China, India and Brazil, added his voice to the impending sense of crisis. 'We go to Durban with no illusion at all that it will be a walk in the park', Zuma said. 'For most people in the developing countries and Africa, climate change is a matter of life and death.'[15]

He was joined by Ban Ki-moon who pleaded with developed countries to spare the Kyoto Protocol. 'It would be difficult to overstate the gravity of this moment', Ban said at the start of a four-day meeting of environment ministers. 'Without exaggeration, we can say: the future of our planet is at stake.'[16] The following day, Canada gave its answer. 'Kyoto, for Canada, is in the past', environment minister Peter Kent told the conference.[17]

The EU laid down its demands: all major emitters to back the completion of a legally binding global climate treaty, perhaps by 2015, into which the Protocol could be subsumed. 'That is what a roadmap should do: describe some principles, the process and the timetable for what should come next', declared the EU environment commissioner Connie Hedegaard. 'Without a roadmap, no second commitment period.'[18]

Hedegaard was backed up by Britain's climate and energy secretary Chris Huhne who underlined that the EU would not agree to a second Kyoto commitment period without 'hard, bankable' commitments from other large nations.[19] To act in isolation 'makes no sense', Huhne told the conference. It would not control 85 per cent of global emissions and it would not give green energy investors the certainty they needed.

> The roadmap and the second commitment period are part of the same package, the same route towards a legally binding global deal. They cannot be separated from one another, and we will not let them be.[20]

This was delusory. The Bali Road Map was to have ended with a Copenhagen climate treaty. Countries could not be expected to commit in good faith to the form of an agreement before knowing what was in it.

Like the US, the EU also wanted to knock down the climate change Berlin Wall. 'We need to discuss whether we can continue to divide the world in the traditional thinking of the North and the South, where the North has to commit to a binding form whereas the South will only have to commit in voluntary form', Hedegaard told reporters. Here she ran into opposition from the new convention executive secretary, Christiana Figueres. 'The North-South divide over historical responsibility still has more weight than the forward-looking approach of respective capabilities', Figueres responded, demonstrating where the convention secretariat's sympathies lay.[21]

Momentum towards a new treaty put US negotiators on the spot. When it came to his turn to address the conference, Stern was heckled by an American student. 'We need an urgent path to a fair, ambitious and legally binding treaty', Abigail Borah told the conference to sustained applause before being bundled out of the hall. When he spoke, Stern denied that the US had been dragging its feet or postponing action until after 2020. By noting that the EU had called for a road map 'that the US supports', the *New York Times* reported that Stern had endorsed the move toward a treaty 'somewhat ambiguously'.[22]

The EU got assistance from an unexpected quarter. Normally sure-footed negotiators, the Chinese delegation undercut their position— and that of the US—by indicating that China might be willing to accept a legally binding post-2020 treaty. The momentum towards a treaty appeared irresistible. As the COP went into overtime, there was a lone hold-out.

'India will never be intimidated by threats', Jayanthi Natarajan, India's chief negotiator, said. 'How do I give a blank cheque and give a legally-binding agreement to sign away the rights of 1.2 billion people?'[23] In the spirit of a Zulu *ubuntu*, Natarajan and Hedegaard went into a huddle and emerged with more words added to the text on the proposed destinations of the road map. 'Legal framework' had been ditched for 'protocol or legal instrument'. When 'legal outcome' was inserted at the last minute, the Europeans threatened not to endorse the proposals. India, supported by China, strongly objected to its deletion. At the suggestion of the chief Brazilian negotiator, it was replaced by 'agreed outcome with legal force'. Thus the COP decided:

> to launch a process to develop a protocol, another legal instrument or an agreed outcome with legal force under the Convention applicable to all parties.[24]

Durban was a triumph for EU climate diplomacy. For the first time in two decades of climate change negotiations, the EU had pulled apart the G-77. It then detached Brazil and South Africa from the BASIC four, helped by China's apparent abandonment of a long-standing position. As Stern drily observed, 'this is a significant package, I think, a very significant package.'[25] It wasn't to last.

At the 2012 COP in Doha, it was back to business as usual. With Beijing insisting on the division between developed and developing nations, Stern said the next climate deal must be based on 'real world' considerations, not 'an ideology that says we're going to draw a line down the middle of the world.'[26] Stern's remarks drew a swift response in an unsigned attack in the Xinhua News Agency showing the flimsiness of the Durban Platform and just how little had changed.

> As usual, the United States has challenged the principle of "common but differentiated responsibilities" … saying that the future agreement on coping with climate change should be based on "real world" considerations and it should not specify different responsibilities for rich and poor countries. But this really depends on what kind of real world the US is living in. For 1.3 billion Chinese, the world is made up of developing and developed countries in which people live very different lifestyles and are capable of doing different things … Between developed and developing nations, there is a world of difference. That's why equality can only be realised when different players bear obligations in line with their capacities.[27]

Discussions on agreeing to a shared vision got bogged down, with the EU complaining that the talks had yielded 'no progress' and the US opposing references to equity and 'common but differentiated responsibilities'—a formula viewed by developing nations as the bedrock principle of the 1992 climate change convention, though Stern later backed off.[28] They were backed by the UN secretary general. 'The climate change phenomenon has been caused by the industrialisation of the developed world',

Ban said in an interview. 'It's only fair and reasonable that the developed world should bear most of the responsibility.' Hedegaard hit back. It was clear that rich countries should do more than poorer countries, the EU commissioner said, 'but all of us will have to do the maximum we can.'[29]

China tried to insert language that backtracked from the Durban agreement on all countries taking binding action. 'We're doing ridiculous things', Chinese delegate Su Wei said, before withdrawing—for the time being.[30] Still, Hedegaard remained optimistic:

> We are crossing the bridge from the old climate system to the new system.
> Now we are on our way to the 2015 global deal.[31]

Instead the COP became embroiled in arguments about money. Developing nations were concerned about the lack of detail on ramping up of climate cash to the 2020 level. 'It's troubling that some developing countries, for example the US, are very sceptical toward doing anything beyond saying "we have made a promise of $100bn by 2020"', said Baard Sohjell, environment minister of oil-rich Norway.[32] The reason was not hard to find. As an EU climate change negotiator explained, 'these are tough financial times in Europe.'[33]

However the parties did agree that the next COP in Warsaw would create 'institutional arrangements' to compensate countries for loss and damages caused by climate change. 'This is a historic decision because it ends a twenty-year discussion on if and how loss and damage from climate change will be addressed', Farhana Yamin, an environmental lawyer and former adviser to Hedegaard, said. According to Greenpeace, it 'finally puts the climate change bill on to the table at the UN talks.'[34]

If Durban was the zenith of the post-Copenhagen negotiations, the November 2013 Warsaw COP was its nadir. The government of Shinzo Abe tore up Japan's Copenhagen Accord commitment to cut its emissions by 25 per cent compared to 1990 levels, replacing it with a new target that implied a 3.8 per cent increase. 'I don't have any words to describe my dismay', China's Su Wei told reporters. 'This is not only a backward movement from the Kyoto Protocol, but also a startling backward move from the Convention'—a crude exaggeration illustrating China's intent to exploit Japan's move.[35] It also showed that, other than the EU with its regime of self-binding targets, the moral and political commitments of the Copenhagen accord—the approach championed by the US—weren't worth the paper they were written on.

Adding fuel to the flames, as it were, Poland—the world's ninth largest coal mining nation—hosted a summit of coal producers not far from the COP. Christiana Figueres gave the miners a pep talk. 'I am here to say that coal must change rapidly and dramatically for everyone's sake', Figueres told them, though she said this didn't require the immediate disappearance of coal.[36]

Back at the COP, there was turmoil after Brazil resurrected its idea of creating a formula to calculate historical blame that it had first proposed in 1997 in the run-up to Kyoto. 'They must know how much they are actually responsible', said Brazil's Raphael Azeredo.[37] Few things were better calculated to widen the gulf between developed and developing nations.

It got worse. Connie Hedegaard accused a group of 'like-minded developing countries' (a new grouping with China and India at its core) of opposing a push to the 2015 deal by insisting on the rich-poor country firewall. 'It is not acceptable to the European Union', she said. Venezuela, speaking on behalf of the group, called the commissioner's comments a 'brazen attack', responsible for seriously damaging the atmosphere of confidence and trust.[38] By then, it was open season on Durban. China requested a reference to an article in the 1992 convention to say that only developed countries are required to cut their emissions. The move alarmed Todd Stern. 'I hope I'm wrong about what I heard, but it would certainly be disappointing to move backward in time, not forward toward Paris,' Stern commented.[39]

The worst came last. Negotiators wrangled over replacing 'commitment' with 'contribution' in the draft text on advancing the Durban Platform. Su Wei explained the significance of the change. 'Only developed countries should have commitment', he told the conference. Emerging economies could merely be expected to 'enhance action'.[40] India's Jayanthi Natarajan put it brutally: 'The firewall exists and it will continue to exist'.[41]

The final wording of the COP decision demonstrates the extent of Durban's disembowelment, asking parties to make preparations:

> for their nationally determined contributions, without prejudice to the legal nature of the contributions, in the context of adopting a protocol, another legal instrument or an agreed outcome with legal force … [42]

To all intents and purposes, the Durban Platform had been demolished and the bankable promises, on which the EU relied on to extend its Kyoto commitment, turned out to be duds.

The day after the Warsaw COP ended, China's lead negotiator Xie Zhenhua spelled out what it meant. 'Contributions' is a neutral word, which can be interpreted as either 'commitments' made by developed countries or 'actions' taken by developing ones.

Since Copenhagen, four rounds of climate conferences—interspersed with numerous non-COP sessions—had achieved nothing substantive other than establish the principle of climate change loss and damages, potentially opening the door to a bonanza for tort lawyers. The EU and the US under the Obama administration were the two parties with the strongest interest in obtaining a positive outcome from the negotiations. Yet the Europeans dismissed the gimlet-eyed realism of US negotiators and their pragmatic view that getting something agreed was better than nothing—an approach that was driven off the field by the temporary success of the Europeans at the Durban COP.

The Europeans were consumed by the desire to turn the Paris climate change summit in 2015 into an all-or-nothing replay of Copenhagen six years earlier. Nothing had happened to change the fundamental interests of the key players and therefore deliver a different outcome. In one respect, though, the Europeans got what they wanted. The false dawn they created at Durban was sufficient for the EU, together with the rest of Europe, to hoodwink itself into signing up to a second phase of Kyoto just as all the other Western nations were queuing up at the Kyoto exit door to join the US. The EU had invested too much political capital in global warming and too much physical capital in decarbonisation for it to be able to pull the plug on Kyoto.

If the cost of cutting carbon dioxide emissions was relatively modest, doubtless there would have been a deal done years ago. At Durban, Christiana Figueres characterised the discussions as being about 'nothing short of the most compelling energy, industrial, behavioural revolution that humanity has ever seen.'[43] To expect nearly 200 independent nations to voluntarily and irrevocably commit, as a conscious act of policy, to such unprecedented economic upheaval always was a pipedream. So it has proved.

14   The hockey stick: a retrospective

Ross McKitrick

The fact that times in the past experienced a warmer climate than today is highly inconvenient for the proponents of the theory of human induced global warming. So the finding in 1998 that temperature trends were much higher in the present day than the past by the American climatologist Michael E. Mann was a key part of the political move towards climate change policy.

Rather than the climate oscillating between hotter periods (for instance, the medieval warm period) and cooler periods (for instance, the little ice age between the fourteenth and nineteenth centuries), Mann's findings suggested that global warming was out of control. The graph of temperatures in the northern hemisphere looked less like a wave, and more like a 'hockey stick'.

Mann's redrawn temperature graph was in part based on a study of Siberian tree rings. Ross McKitrick and Steve McIntyre studied the data behind the evidence and the statistical techniques used to interpolate it and found key errors in the statistical technique used to combine Mann et al.'s blended data would almost invariably produce a 'hockey stick'.

The controversy led to a US Senate Committee setting up an inquiry under Professor Edward Wegman, who essentially reported that Mann's claims—that the 1990s were the hottest decade in 1000 years—could not be supported. We asked Ross McKitrick to write this synopsis of the issue.

—Alan Moran, Editor

The best place to start when learning about the hockey stick is Andrew Montford's superb book *The Hockey Stick Illusion*.[1] Other essential sources are the original Mann et al. papers,[2] the McIntyre and McKitrick papers,[3] Steve McIntyre's and my presentation to the National Academy of Sciences Panel,[4] McIntyre's Ohio State University presentation,[5] a few survey papers and chapters of mine,[6] and McIntyre's *climateaudit. org* posts over the past decade on proxy quality, the Yamal substitution, the Briffa truncation, data secrecy, and some other issues.[7]

It is sometimes said that we found Michael Mann's algorithm would always produce a hockey stick, even from random numbers (Figure 2). That is not quite right: we found that the algorithm *could* do so, given the right kind of random numbers (autocorrelated, rather than independent). We also found that it mined for hockey stick shapes and overstated their dominance in the underlying data patterns, and that it understated the uncertainties of the resulting climate reconstruction (or equivalently, exaggerated the significance).

Figure 1: Temperature chart as it appeared in the 1990 IPCC report

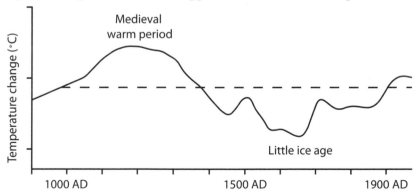

Source: C. K. Folland, R. T. Karl, K. YA. Vinnikov, "Observed Climate Variations and Change," Climate Change: The IPCC Scientific Assessment, ed. J. T. Houghton, G. J. Jenkins and J. J. Ephraums (Cambridge: Cambridge University Press, 2010), 202, accessed July 15, 2014, http://www. ipcc.ch/ipccreports/far/wg_I/ipcc_far_wg_I_chapter_07.pdf, chart 7.1, page 202.

Figure 2: Northern Hemisphere temperature chart as it appeared in the 2001 IPCC report (the 'Hockey Stick' graph)

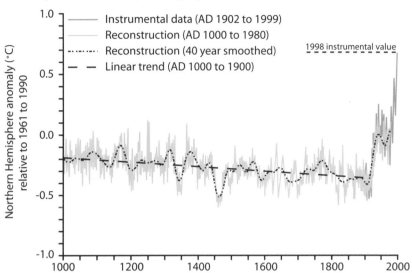

Source: IPCC, "Summary for Policymakers: Report of Working Group I of the Intergovernmental Panel on Climate Change," in Climate Change 2001: The Scientific Basis: Contribution of Working Group I to the Third Assessment Report of the Intergovernmental Panel on Climate Change, ed. J. T. Houghton, Y. Ding, D. H. Griggs, M. Noguer, P. J. van der Linden, X. Dai, K. Maskell and C. A. Johnson (Cambridge: Cambridge University Press, 2001), accessed July 15, 2014, http://www.grida.no/publications/other/ipcc_tar/?src=/climate/ipcc_tar/wg1/005.htm figure 2.20

A very brief summary of the problems of the hockey stick would go like this. Mann's algorithm, applied to a large proxy data set, extracted the shape associated with one small and controversial subset of the tree rings records, namely the bristlecone pine cores from high and arid mountains in the US southwest. The trees are extremely long-lived, but grow in highly contorted shapes as bark dies back to a single twisted strip. The scientists who published the data had specifically warned that the ring widths should not be used for temperature reconstruction, and in particular their twentieth century portion is unlike the climatic history of the region, and is probably biased by other factors.[8]

Mann's method exaggerated the significance of the bristlecones so as to make their chronology out to be the dominant global climatic pattern rather than a minor (and likely inaccurate) regional one; Mann then understated the uncertainties of the final climate reconstruction, leading to the claim that 1998 was the warmest year of the last millennium, a claim that was not, in reality, supportable in the data. Furthermore, Mann put obstacles in place for subsequent researchers wanting to obtain his data and replicate his methodologies, most of which were only resolved by the interventions of US Congressional investigators and the editors of *Nature* magazine, both of whom demanded full release of his data and methodologies some six years after publication of his original *Nature* paper.

Mann had re-done his hockey stick graph at some point during its preparation with the dubious bristlecone records excluded and saw that the result lost the hockey stick shape altogether, collapsing into a heap of trendless noise. However he never pointed this out to readers.

He also stated that he had computed test scores called r^2 statistics that he said (or implied) confirmed the statistical significance of his results, yet when the r^2 scores were later revealed they showed no such thing, and by then he had taken to denying he had even calculated them.

Our critique of Mann's method

There are two key parts to the hockey stick-making machine. The first is the principal components (PC) step, and the second is the least squares (LS) fitting step. The PC step takes large numbers of temperature proxies and compiles them into a relatively small number of composite series. The LS step then lines up the final segment of the composites against an upward-sloping temperature graph and puts weight on them in proportion to how well they correlate. If there are many composites and only one has a hockey stick shape, the LS step will find it and put most of the weight on it. If none of the composites has a hockey stick shape, then the LS step will come up blank and the resulting graph will just look like noise.

Mann's PC step was programmed incorrectly and created two weird effects in how it handled data. First, if the underlying data set was mostly random noise, but there was one hockey stick-shaped series in the group, the flawed PC step would isolate it out, generate a hockey stick composite and call it the dominant pattern, even if it was just a minor background fluctuation. Second, if the underlying data consisted of a particular type of randomness called 'red noise'—basically randomness operating on a slow, cyclical scale—then the PC step would rearrange the red noise into a hockey stick-shaped composite. Either way, the resulting composites would have a hockey stick shape for the LS step to glom onto and produce the famous final result.

The use of red noise series is necessary for testing the statistical robustness of the hockey stick method. This is a procedure called Monte Carlo analysis. For one of our 2005 papers, we generated thousands of series of trendless autocorrelated random numbers and ran them through the PS and LS steps.[9] This generated thousands of results, each of which had an index of accuracy called the reduction of error (RE) score. Likewise the actual proxy data has an associated RE score. We set a benchmark based on the idea that, if the proxy data was actually informative about the real world it had to yield a higher RE score than most of the (uninformative) artificial data. Mann had done the same thing, but had not taken into account the effect of the erroneous PC method. The real proxy data didn't turn out to be more informative than red noise, but he set his benchmark too low, making his proxy results look statistically significant when in reality they weren't.

There was a big red flag in his calculations that should have tipped him off. Another model test is called the r^2 score. It has the nice feature that you don't need to do Monte Carlo simulations, it has standard benchmark tables available in any statistics textbook.[10]

While Mann reported the (favourable) r^2 scores for the later portion of his graph,[11] he didn't mention them for the early portion (pre-1750), where they were nearly zero, indicating a lack of statistical significance. Instead he only reported the RE score, which he thought indicated significance. He showed the reader the RE test that he thought (incorrectly) was favourable, yet he kept referring to significance *tests* in the plural in support of his claims, so the reader would naturally assume the unreported r^2 scores looked good too.[12]

They didn't, but he failed to report that in the article. And as we later showed, the r^2 and RE scores were actually saying the same thing, namely that the hockey stick was uninformative as an indicator of past temperatures.

Stickhandling

In 2005, following an article on the dispute in *The Wall Street Journal*, Mann had been sent a list of questions by the Energy and Commerce Committee of the US Congress, one of which was whether he had computed the r^2 score. His answer was:

> My colleagues and I did not rely on this statistic in our assessments of "skill" (i.e., the reliability of a statistical model, based on the ability of a statistical model to match data not used in constructing the model) because, in our view, and in the view of other reputable scientists in the field, it is not an adequate measure of "skill." The statistic used by Mann et al. 1998, the reduction of error, or "RE" statistic, is generally favored by scientists in the field.[13]

The answer is classic misdirection. He was not asked: 'Did you rely on the r^2 score when assessing your results?' There was no need to ask that: if he *had* relied on it he would never have claimed his results were significant. He only claimed significance by *ignoring* it. The question specifically was whether he computed r^2. Tellingly, in his reply he changed the subject. But it hardly matters. Either he did not compute it, in which case he was lying in the paper by saying he had, or he did, in which case his failure to disclose it was misleading to his readers.

When we and Mann appeared before the National Academy of Sciences (NAS) panel in 2006, we presented this issue in detail. We showed the panel that the Supplementary Information to the Mann et al. 1998 paper did not report the verification r^2 scores, and we urged them to ask Mann whether he had computed them. John Christy of the University of Alabama-Huntsville put the question to him. To our astonishment, Mann point-blank denied having done so, claiming it would be 'silly and incorrect reasoning.'[14] Mann then launched an extraordinary tirade against r^2, a well-understood statistic which is found in every statistics textbook and is the workhorse of model testing.

After the NAS panel hearings we wrote a letter to the chair, Gerald North, expressing our frustration that they allowed Mann to get away with this, but we were not successful in getting them to follow up on the matter.[15]

The National Academy of Sciences Report

More evidence that the Mann procedure exaggerated the statistical significance of his results came in the NAS Panel Report itself.[16] While they lost the plot on the r^2 issue, they did at least look at the overall question of how to assess a statistical climate reconstruction. They came up with the most elliptical way possible to say that the Mann hockey stick was unreliable. Here is what they said:

> Reconstructions that have poor validation statistics (i.e., low CE) will have correspondingly wide uncertainty bounds, and so can be seen to be unreliable in an objective way. Moreover, a CE statistic close to zero or negative suggests that the reconstruction is no better than the mean, and so its skill for time averages shorter than the validation period will be low. Some recent results reported in Table 1S of Wahl and Ammann[17] indicate that their reconstruction, which uses the same procedure and full set of prox-

ies used by Mann et al ... gives CE values ranging from 0.103 to −0.215, depending on how far back in time the reconstruction is carried.[18]

Sir Humphrey Appleby could not have phrased it better. Unpeeling the obfuscations, here is what they said:

- Reconstructions can be assessed using a variety of tests, including RE, r^2 and the coefficient of efficiency (CE) scores;
- If the CE score is near zero or negative your model is junk;
- Wahl and Ammann include a table in which they use Mann's data and code and compute the test scores that he didn't report; and
- The CE scores range from near zero to negative, which tells us that Mann's results were junk.

Another exercise in obfuscation concerned the reliance on bristlecones. The NAS report said the following:

Such trees [bristlecones] are sensitive to higher atmospheric CO_2 concentrations ... possibly because of greater water-use efficiency ... or different carbon partitioning among tree parts ... While 'strip-bark' samples should be avoided for temperature reconstructions, attention should also be paid to the confounding effects of anthropogenic nitrogen deposition ... For periods prior to the 16th century, the Mann ... reconstruction that uses this particular principal component analysis technique is strongly dependent on data from the Great Basin region in the western United States. Such issues of robustness need to be taken into account in estimates of statistical uncertainties.[19]

Stripping away the bark, here is what this means:

- Bristlecone records are sensitive to a variety of environmental conditions other than temperature and should be avoided for climate reconstructions;
- Mann's results strongly depend on the bristlecone records; and
- His results are therefore not robust, an important point over and above the lack of statistical significance.

The NAS report also made a few other points, buried in elliptical prose or scattered around the report where the press would be sure of never finding them (not that they looked). Putting them together, they upheld all the claims in our submission:

> McIntyre and McKitrick[20] demonstrated that under some conditions, the leading principal component can exhibit a spurious trend-like appearance, which could then lead to a spurious trend in the proxy-based reconstruction …[21] As part of their statistical methods, Mann et al. used a type of principal component analysis that tends to bias the shape of the reconstructions.[22]

The report even included its own graphical replication of the artificial hockey stick effect from feeding red noise into Mann's algorithm, and noted that the usual RE significance benchmark 'is not appropriate'[23] and that 'uncertainties of the published reconstructions have been underestimated'.[24]

The censored folder

Mann also published an online review article in 2000 that assured readers in categorical terms that their results were 'robust' to non-climatic bias in tree ring data[25] and even to the complete removal of tree rings from their data set, though they illustrated that point only for the post-1760 interval.[26] In the course of our analysis, McIntyre found some directories at Mann's FTP site (the 'CENSORED' directories), which, through detective work, were found to contain assessments of the impact from dropping the bristlecones from the underlying data. In light of the claim in Mann et al.,[27] this should not have made any difference, but it did. In our NAS presentation we showed graphs of the data in Mann's 'CENSORED' results, in which the hockey stick shape completely disappears. That is, even applying Mann's biased methods, after dropping the few bristlecone pine series there is no remaining hockey stick shape. The claim in Mann et al. about robustness to the exclusion of the tree ring data was obviously misleading.

In the letter from the Congressional Oversight committee to Mann he was asked:

> Did you run calculations without the bristlecone pine series referenced in the article and, if so, what was the result?[28]

Mann's answer was lengthy, but included the following:

> For a complete scientific response, you should consult the article my co-authors and I published back in 1999 addressing precisely these issues … As my co-authors and I explained … given the proxy data available at that time, certain key tree-ring data … were essential, if the reconstructed temperature record during early centuries were to have any climatologic "skill" (that is, any validity or meaningfulness). These conclusions were of course reached through analyses in which these key datasets were excluded, and the results tested for statistical validity. Our conclusions have been confirmed by Wahl and Ammann.[29]

Translation: Yes. When we removed them the graph collapsed and the statistical scores went to zero. Oh dear, didn't we mention that? Anyway, to avoid the problem, we kept them in.

Mann's claim that the 1999 paper addressed 'precisely these issues' was misleading. In that paper they did mention that their top-weighted PC was 'essential' but they didn't report the results of excluding the bristlecones.[30] Instead they applied a 'correction' that they claimed (without proof) fixed the contamination pattern in the bristlecones, even though it only applied to the nineteenth century portion. And their 2000 paper claimed robustness both to contamination of bristlecones and removal of tree ring data.[31] Wahl and Ammann later offered the argument that since the hockey stick fails all statistical tests without the bristlecones they ought to be retained (the logic really was that bad).[32]

In our letter to North we pointed out that we agreed with Wahl and Ammann (and Mann) that the reconstruction without the bristlecones is no good.

But, we added:

> Our contention is that the reconstruction *with* bristlecones is also no good, as evidenced by the failure of verification r^2 and CE statistics.[33]

Conclusion

The story continued on from there and much more could be said. The intensity with which so many people have followed the story, and its continuing relevance via the ongoing *Mann v. Steyn* lawsuit (as well as others), indicate to me that it is more than just an academic spat about proxy quality and r^2 scores. I suspect that the whole episode has wider social significance as an indicator of a rather defective aspect of early twenty-first century scientific culture.

15   The IPCC and the Peace Prize

Donna Laframboise

In late 2012, the art director of a reputable magazine contacted photographer Alex Waterhouse-Hayward. How would he like to meet a Nobel laureate?[1] Three months later, the photographer posted online a black-and-white portrait, announcing to the world that this was 'Nobel Laureate Mark Jaccard.'

But Jaccard has never won a Nobel Prize. Visit NobelPrize.org, type his name into the search box, and you'll find no mention of him. Why did the smart people at Canada's *The Walrus* magazine think otherwise? Why did the cover of this award-winning publication wrongly describe him as a 'Nobel economist?'

The short answer is that Jaccard is one of an estimated 9,000 people who've helped the Intergovernmental Panel on Climate Change (IPCC) write its many reports over a span of 25 years. In late 2007, following news that this UN body had been awarded the Peace Prize jointly with Al Gore, the IPCC's chairman profoundly over-stepped his authority. Writing to IPCC-affiliated academics *en masse*, Rajendra Pachauri mistakenly proclaimed 'This makes each of you Nobel Laureates.'

Everyone should have understood that this was mere rhetorical flourish. When an individual wins a Nobel, they are contacted directly by Nobel officials and later receive a significant sum of money. As Australian researcher John McLean would later tell New Zealand meteorologist Kevin Trenberth, 'Pachauri can't hand-out laureates like cups of coffee, and you, Kevin, surprise me by seeming to believe that he can.'

The IPCC isn't the first organisation to be honoured in this manner. In 1977, the Peace Prize went to Amnesty International. In 1999, it was Doctors Without Borders. In between, in 1988, the efforts of UN peacekeeping forces were recognised.

If someone who once served in a peacekeeping capacity described themselves as a Nobel laureate, we'd have no trouble identifying them as an insecure ego run amok. But the climate world isn't a normal one. Instead, it resembles the Wild West. Poorly socialised adolescents swagger and bluster, grownups are in short supply, and the sheriffs turn out to be as lawless as everyone else.

The Walrus misled its readers because a significant part of the climate community chose to embrace a Nobel fiction. The unadorned truth was door number one. Cringe-worthy exaggeration was door number two. Many IPCC personnel—the very people entrusted to give us the straight goods about climate change—made the wrong call.

Jaccard was among 23 individuals who worked on one chapter (out of 47) of the IPCC's 1995 report. Along with 24 others, he helped write a second chapter. The role he played was so minor that, when he co-authored a 2007 book on climate change with a Canadian journalist, his IPCC involvement wasn't even mentioned. Shortly afterward, however, he suddenly became a Nobelist.

In 2008, an activist group issued a press release about a report written by Jaccard. The release described him as 'a winner of the 2007 Nobel Peace Prize'. In 2009, a poster advertising an event at a public library featured his photograph alongside large text that identified him as an 'author and Nobel Laureate'.

When he submitted written testimony to a regulatory agency in 2011, Jaccard said he'd 'been honoured with the Nobel Peace Prize' as an IPCC contributor. His employer, Canada's Simon Fraser University, has told the world that he is a Nobel laureate. So have journalists.

Because the IPCC is an international body, there are many Jaccards in many countries. University of Melbourne meteorologist David Karoly has similarly been described by the Australian Broadcasting Corporation as a 'Nobel prize-winning scientist.'[2] Notice the absence of the crucial word 'peace' in that phrase. The problem isn't merely that Karoly has no valid claim to that honour. It's that, when we're told that a *scientist* has won a Nobel, most us of us don't think of soft focus Peace Prizes.

The public is being urged to pay special attention to Karoly's views because he's a Nobel laureate. But Peace Prizes are not endorsements of a person's scientific acumen. They are conferred by a different body than the one that awards physics Nobels. Moreover, the IPCC shared its Peace Prize with a politician—Al Gore—whose dismal grades in university-level science courses are a matter of public record.[3]

For years, Karoly has delivered presentations with titles along the lines of 'Lies, Damn Lies and Climate Change Skeptics'. In a podcast available on iTunes, he discusses 'the lies that are told in the [climate] debate'.

It's awkward, therefore, to discover that the Science Club of Melbourne High School, an elite all-boys academy, described Karoly as 'a Nobel prize winner' when talking about his May 2014 visit—and that the website of the Australian Medical Students' Association continues to falsely tell the world he was 'awarded the 2007 Nobel Peace Prize.'[4] More awkward still is the Victorian church group, the Church of All Nations, whose website makes that same false claim.[5] Then there was the event poster for a fundraiser associated with the Yarra Climate Action Now Community Solar Project. It wrongly described the host as 'climate scientist and Nobel Peace Prize winner' Karoly.

Elsewhere, Julie Arblaster, now an employee of Australia's Bureau of Meteorology (and a former Karoly-supervised PhD student), declares on her CV that she became a 'Nobel Peace Prize laureate' six years prior to completing her 2013 doctorate.[6]

On its website, the government of New South Wales continues to incorrectly advise the public that Mark Howden—an agricultural specialist and career public servant—is a 'Nobel Peace Prize winner' due to his IPCC affiliation.[7] Similarly, in a Brisbane Meeting Planners' guide for 2011-2013, Brisbane City Council boasted that a keynote address by 'Nobel Prize winner' Howden was one of the high points of a convention held recently in that city.[8]

The scale of this misunderstanding is unlikely to have extended so far had the chairman of the IPCC not been setting the worst possible example. The fact that Rajendra Pachauri attended a ceremony and delivered an acceptance speech on behalf of his organisation did not magically transform him, personally, into a Nobel laureate. There should have been no confusion on this point since the prize money wasn't given to Pachauri—it was bestowed on the IPCC, which used it to fund scholarships.

But a startling array of respectable organisations apparently believe that Pachauri himself became a Nobelist via the same application of pixie dust that transformed other IPCC personnel. The first thing Amazon.com tells us about his 2010 novel, *Return to Almora*, is that it was written by a 'Nobel laureate'.[9]

Take a spin with an internet search engine and thousands of news articles, interviews, videos, and photos all promulgating the myth that Pachauri has personally received a Nobel prize will present themselves. Again and again, the implication is that the public should pay close attention to what Pachauri is saying. He isn't just anyone, we're told—he's a Nobel laureate. Except that he's not.

Among those who've erroneously described Pachauri in this manner we find: the US Secretary of State, the office of the Prime Minister of Norway, the Mayor of London, the New York Academy of Sciences, the United Nations Environment Programme, the UN Economic Commission for Africa, United Nations Children's Fund (UNICEF), the World Bank, the Asian Institute of Technology, the World Wildlife Fund, *350.org*, *EarthDay.org*, *DemocracyNow.org*, the Swiss Broadcasting Corporation, the Australian Broadcasting Corporation, the BBC, *The New York Times*, *The Japan Times*, *Der Spiegel*, *The Vancouver Sun*, and *The Times of India*.

This fiction has also infected university campuses. In 2009, North Carolina State University issued a press release titled 'Nobel Laureate to Deliver Fall Commencement Address at NC State'. Judging by its first sentence—which describes Pachauri as a Nobel Prize winner and NC State alumnus—everyone who attended that event was misinformed in this regard.

Yale University's *Environment 360* magazine, a prestigious, specialist publication that should know better, titled an article 'A Conversation with Nobel Prize Winner Rajendra Pachauri'. A few years ago, Utrecht University, in the Netherlands, sponsored a lecture by 'Nobel Laureate Rajendra Pachauri' as part of a masters degree program. The University of Eastern Finland has gone so far as to make 'Nobel Prize Winner, IPCC Chair Dr Pachauri' its first honourary professor.

When Deakin University, Australia's ninth largest university, announced the opening of a research centre in cooperation with Pachauri's The Energy and Resources Institute, it told the world it was partnering with a 'Nobel Prize winner'. And then there's that list on the website of Gustavus Adolphus College in Minnesota. It enumerates the Nobel laureates on whom this institution has also bestowed an honorary doctorate. Currently, Pachauri's name appears first.

To recap: back in 2007, Pachauri elevated thousands of people, including himself, to the status of Nobel laureates. And no one called him on it. Not the Nobel committee. Nor the UN bodies that established the IPCC. Nor the media. Did we all lose our minds?

The world's science academies deserve special scrutiny in this regard. Why have they remained mute, year after year, as Pachauri and others are falsely described as Nobel laureates? If the public can't count on science academies to police a matter this straightforward, what purpose do they serve?

Eventually, however, hot air deflates. In October 2012, American meteorologist Michael Mann filed a defamation lawsuit against two journalists and two publishers. The second paragraph of his 37-page legal document reads as follows:

> Dr. Mann is a climate scientist whose research has focused on global warming. Along with other researchers, he was one of the first to document the steady rise in surface temperatures during the 20th Century and the steep increase in measured temperatures since the 1950s. *As a result of this research, Dr. Mann and his colleagues were awarded the Nobel Peace Prize.*[10] [emphasis added]

Notice there isn't any mention of the actual recipient of that award—the IPCC. Someone unfamiliar with this matter could be forgiven for thinking that Mann himself was the person at the Nobel podium. On page five, we learn that Mann served as a lead author on a single chapter of the IPCC's 2001 report. On the next, this document implies once again that the prize is connected to Mann's own research:

> The work of Dr. Mann and the IPCC has received considerable accolades within the scientific community. In 2007, Dr. Mann shared the Nobel Peace Prize with the other IPCC authors for their work in climate change, including the development of the Hockey Stick Graph.[11]

But the reasoning behind this award was rather different. The Peace Prize committee stated clearly that it believes climate change 'will increase the danger of war'. The prize therefore recognised the efforts of Al Gore and the IPCC to educate the world 'concerning man-made climate changes and the steps that need to be taken to counteract those changes.'

To claim, in the context of legal proceedings, that you received a Nobel Prize for contributing to one chapter in one UN report is absurd. But Mann's legal document goes further, childishly suggesting that this *faux* honour should insulate him from criticism:

> It is one thing to engage in discussion about debatable topics. It is quite another to attempt to discredit consistently validated scientific research through the professional and personal *defamation of a Nobel prize recipient.*[12] [emphasis added]

Apparently, you're being extra naughty if you defame someone who's won a Nobel.

This is the poisoned fruit that Pachauri's pixie dust summoned into existence. When he told thousands of people they were Nobel laureates, the poorly socialised adolescents took him seriously. They started to believe that their own research had received this honour. Henceforth, anyone challenging their work was an anti-science moron.

After examining Mann's legal document, a journalist contacted a Nobel official in an attempt to confirm that Mann was, indeed, the recipient of a Peace Prize. The official said he was not. Presumably, someone from the Nobel organisation then quietly suggested to the IPCC that it would be a good idea to clarify this matter.

Two weeks after Mann filed his legal papers, the IPCC issued a one-page 'Statement about the 2007 Nobel Peace Prize' that flatly contradicts Pachauri's 'This makes each of you Nobel Laureates' claim. It reads, in part:

> The prize was awarded to the IPCC as an organization, and not to any individual associated with the IPCC. Thus it is incorrect to refer to any IPCC official, or scientist who worked on IPCC reports, as a Nobel laureate or Nobel Prize winner.[13]

The IPCC posted the above statement on its website, but it did not send a copy to the same list of people who received Pachauri's erroneous proclamation five years earlier. Nor did it issue a press release.

The internet is currently saturated with accounts that falsely describe Pachauri and other IPCC personnel as Nobel laureates. Numerous academics continue to make inaccurate claims about their relationship to this prize in their CVs and other biographical material. But setting the record straight is not an IPCC priority.

Media outlets aren't known for being overly concerned about their own mistakes. When *The Walrus* magazine realised it had erroneously described Jaccard as a Nobel laureate (on its cover, in a headline, as well as in its table of contents), did it issue a proper *mea culpa*? Did it tell its readers about the international scale of the misinformation it had inadvertently helped to promulgate?

I'm afraid not. Jaccard's article appeared in the March 2013 print edition. Three issues later, in June 2013, the letters-to-the-editor section of *The Walrus* published the remarks of eleven individuals over two pages. Item twelve, appearing at the very end of that section, read as follows:

Tusk-Tusk

The March issue of *The Walrus* identified Mark Jaccard as a Nobel laureate. While he was part of the Intergovernmental Panel on Climate Change, which won the Nobel Peace Prize in 2007, individual members are not recognized as laureates.[14]

Tusk-tusk. When the magazine was employing the term 'Nobel' on its cover, that word had heft and meaning. Three months later, it became a triviality—the magazine's misstep didn't even rate the standard 'we regret this error' declaration.

The Walrus' fact-checker should have stopped the Nobel language in its tracks prior to publication. But it isn't difficult to understand how this mistake originally occurred. If they took the time to double-check every statement they encounter, journalists would never get anything written. We all take a great deal on trust, every single day. It doesn't cross the minds of most of us that people who call themselves scientists would fudge something this important.

Unlike marketers, scientists are not supposed to embellish. They're supposed to be clear-eyed about what is true and what is false. The idea that hundreds of scientists have been padding their resumés, the idea that they've been walking around in broad daylight improperly claiming to be Nobel laureates isn't something any normal person would expect.

That said, there's no escaping the fact that this is a story of ongoing journalistic failure. Five years after receiving the Peace Prize, the IPCC finally acknowledged that its chairman and others had been basking in a glory that they did not personally earn. It acknowledged that the medals on their chests were made of tinfoil. A journalist helped prod them into this admission.[15] But it wasn't a prominent journalist from a large media conglomerate who did this. Nor was it a journalist drawing a fulltime salary, with years of experience covering environmental issues. Rather it was a freelancer named Thomas Richard who writes for the online *Boston Environmental Policy Examiner*, and who edits the sceptically-minded ClimateChangeDispatch.com website.

For seven years, much of the international media has misrepresented the connection between the Nobel Peace Prize and IPCC personnel. Journalists have aided and abetted hundreds of self-aggrandising individuals who claimed to be Nobel laureates when they are nothing of the sort. Large swathes of the public have consequently been misled. News accounts, government websites, and institutions of higher learning are now so saturated with this false information that millions of people will continue to be conned for years to come.

The moral of this tale of exaggeration and bamboozlement: on the straightforward question of whether or not certain individuals are Nobel laureates, both scientists and journalists have failed to tell the truth. It is therefore unwise to rely on the word—and judgment—of these same people with respect to whether we face a climate crisis.

16   Global warming's glorious ship of fools
Mark Steyn

Yes, yes—just to get the obligatory 'of courses' out of the way up front: of course 'weather' is not the same as 'climate', and of course the thickest iciest ice on record could well be evidence of 'global warming', just as 40-and-sunny and a 35-below blizzard and twelve degrees and partly cloudy with occasional showers are all apparently manifestations of 'climate change'. And of course the global warm-mongers are entirely sincere in their belief that the massive carbon footprint of their rescue operation can be offset by the planting of wall-to-wall trees the length and breadth of Australia, Britain, America and continental Europe.

But still: you'd have to have a heart as cold and unmovable as Commonwealth Bay ice not to be howling with laughter at the exquisite symbolic perfection of the Australasian Antarctic Expedition (AAE) 'stuck in our own experiment', as they put it. I confess I was hoping it might all drag on a bit longer and the cultists of the ecopalypse would find themselves drawing straws as to which of their number would be first on the roasting spit. On Douglas Mawson's original voyage, he and his surviving comrade wound up having to eat their dogs. I'm not sure there were any on this expedition, so they'd probably have to make do with the *Guardian* reporters. Forced to wait a year to be rescued, Sir Douglas later recalled, 'Several of my toes commenced to blacken and fester near the tips.' Now there's a man who's serious about reducing his footprint.

But alas, eating one's shipmates and watching one's extremities drop off one by one is not a part of today's high-end eco-doom tourism. Instead, the ice-locked warmists uploaded chipper selfies to YouTube, as well as a self-composed New Year sing-along of such hearty un-self-awareness that it enraged even such party-line climate alarmists as Andrew Revkin, the plonkingly earnest enviro-blogger of the *New York Times*. A mere six weeks ago, pumping out the usual boosterism, the Australian Broadcasting Corporation reported that, had Captain Scott picked his team as carefully as Professor Chris Turney, he would have survived. Sadly, we'll never know—although I'll bet Captain Oates would have been doing his 'I am going out, I may be some time' line about eight bars into that New Year number.

Unlike Scott, Amundsen, and Mawson, Professor Turney took his wife and kids along for the ride. And his scientists were outnumbered by wealthy tourists paying top dollar for the privilege of cruising the end of the world. In today's niche-market travel industry, the Antarctic is a veritable Club Dread for upscale ecopalyptics: think globally, cruise icily. The year before the Akademik Shokalskiy set sail as part of Al Gore's 'Living On Thin Ice' campaign (please, no tittering, it's so puerile—every professor of climatology knows that the thickest ice ever is a clear sign of thin ice, because as the oceans warm, glaciers break off the Himalayas and are carried by El Ninja down the Gore Stream past the Cape of Good Horn where they merge into the melting ice sheet, named after the awareness-raising rapper Ice Sheet ...).

Where was I? Oh, yeah. Anyway, as part of his 'Living On Thin Ice' campaign, Al Gore's own luxury Antarctic vessel boasted a line-up of celebrity cruisers unseen since the 1979 season finale of *The Love Boat*—among them the actor Tommy Lee Jones, the pop star Jason Mraz, the airline entrepreneur Sir Richard Branson, the director of *Titanic* James

Cameron, and the Bangladeshi minister of forests Somebody Wossname. If Voyage of the Gored had been a conventional disaster movie like *The Poseidon Adventure*, the Bangladeshi guy would have been the first to drown, leaving only the Nobel-winning climatologist (Miley Cyrus) and the maverick tree-ring researcher (Ben Affleck) to twerk their way through the ice to safety. Instead, and very regrettably, the SS Gore made it safely home, and it fell to Professor Turney's ship to play the role of our generation's Titanic. Unlike the original, this time round the chaps in the first-class staterooms were rooting for the iceberg: as the expedition's marine ecologist Tracy Rogers told the BBC, 'I love it when the ice wins and we don't.' Up to a point. Like James Cameron's Titanic toffs, the warm-mongers stampeded for the first fossil-fuelled choppers off the ice, while the Russian crew were left to go down with the ship, or at any rate sit around playing cards in the hold for another month or two.

But unlike you flying off to visit your Auntie Mabel for a week, it's all absolutely vital and necessary. In the interests of saving the planet, IPCC honcho Rajendra Pachauri demands the introduction of punitive aviation taxes and hotel electricity allowances to deter the masses from travelling, while he flies 300,000 miles a year on official 'business' and research for his recent warmographic novel, in which a climate activist travels the world bedding big-breasted women who are amazed by his sustainable growth. (Seriously: 'He removed his clothes and began to feel Sajni's body, caressing her voluptuous breasts.' But don't worry; every sex scene is peer-reviewed.) No doubt his next one will boast an Antarctic scene: is that an ice core in your pocket or are you just pleased to see me?

The AAE is right: the warm-mongers were indeed 'stuck in our own experiment'. Frozen to their doomsday narrative like Jeff Daniels with his tongue stuck to the ski lift in *Dumb and Dumber*, the Big Climate enforcers will still not brook anyone rocking their boat. In December 2008 Al Gore predicted the 'entire North Polar ice cap will be gone in five years'. That would be December last year. Oh, sure, it's still here, but

223

he got the general trend-line correct, didn't he? Arctic sea ice December 2008: 12.5 million square kilometres; Arctic sea ice December 2013: 12.5 million square kilometres.

Big Climate is slowly being crushed by a hard, icy reality: if you're heading off to university this year, there has been no global warming since before you were in kindergarten. That's to say, the story of the early twenty-first century is that the climate declined to follow the climate 'models'. (Full disclosure: I'm currently being sued by Dr Michael Mann, creator of the most famously alarming graph, the 'hockey stick'.) You would think that might occasion a little circumspection. But instead the cultists up the ante: having evolved from 'global warming' to the more flexible 'climate change', they're now moving on to 'climate collapse'. Total collapse. No climate at all. No sun, no ice. No warm fronts, except for the heaving bosoms in Rajendra Pachauri's bodice-rippers. Nothing except the graphs and charts of 'settled science'. In the Antarctic wastes of your mind, it's easier just to ice yourself in.

17 Cavemen, climate, and computers

Christopher Essex

A debate takes place between a mysterious time traveller and a Paleolithic shaman before an audience of Paleolithics about the merits of cooking with microwave ovens. The mysterious traveller introduces the electromagnetic field and discusses the merits of eating cooked versus raw meat. The shaman sings mystically about what the gods say. Meanwhile the audience chants: 'Club. Club. Club ... '

It sounds bizarre, but I didn't entirely make it up. I have lived it—well, mostly. There wasn't any time travel, of course, and the subject was climate change rather than microwave ovens, but everything else was much the same, modulo some details. Naturally, objections to the Paleolithic aspect may arise: this is the modern world, not the old stone age. Modern people are educated, supporters of science, and not just a bunch of superstitious cavemen, aren't they? Well, let me relate some of my experiences in answer.

Sometimes I give a pop quiz before giving an interview on climate. 'Tell me,' I ask the journalist, 'what you know about the Navier Stokes equations?' It's pretty vague, so nearly any puny, pathetic answer would do. But there is still one wrong answer: 'What're the Navier Stokes equations?' Despite the low standard, they never pass. Some think that's unfair, but is it? Those equations tell us how water and air move—tough to have an intelligent conversation about climate without that.

Scientists are criticised for speaking in incomprehensible ways. But compromising has led to distorted, simplistic gibberish, making the full technicalities appear like a fraud to keep outsiders out. And then the near-gibberish gets set like concrete and called 'science.' But, mathematical equations are a precise language. While little known to popular culture, they're essential to human understanding of nature. Sadly, equations appear freely in public less frequently than pornography, and can prove to be even more socially unacceptable. Have you ever seen a differential equation in a newspaper? Claiming to discuss climate while trying to escape mathematics comes across to me like children wailing for dessert after refusing to eat their vegetables.

But if I forego the quiz and the journalist wants to talk about actual science, doing the interview with scientific technicalities off limits is more challenging than doing charades. At least in charades, when you act out a movie title, you can count on the audience knowing what a movie is. After decades of slavering climate obsession, we are still admonished not to question the persistent and stunningly stunted level of discourse because it's 'unfair.' While many worry about climate, time and again they fail to truly discuss it because of awkward cultural attitudes about science. Instead, I've witnessed something else. It sounds an awful lot like: 'Club. Club. Club … ' Saying it is unfair to question this makes a heartbreaking joke out of us all.

Speaking for the gods

The stranger from the future begins to explain how microwave ovens work. But Maxwell's equations and the dipole moment of the water molecule don't seem to make much headway with the Paleolithics. In desperation, he falls back to the gibberish method. 'Well, the microwaves rub up against the water, causing heat from friction.' Oh, wait. That's the explanation for microwave ovens given to modern humans.

Making fun of us on science is too easy. But given the torrents of prattling nonsense flowing endlessly from the mouths of our illustrious leaders, journalists, and academics on climate, it's richly deserved. I would be sympathetic if I did not believe that people were capable of much better. Most underestimate their scientific side: you can do it human—it's built in. It might seem hard, but like brussel sprouts, differential equations are good for you. Some squirm to avoid their vegetables, and others enable their tortured avoidance. The enablers force us to indulge people who act like cats gingerly stepping around a puddle when scientific technicalities are in play. By all means don't use microwave ovens if you think they infest your brain with leprechauns. However, if we are facing political decisions over policy affecting everyone, you should not get to step gingerly around the puddle. You should either step up and learn, or get out of the way in favour of those who have.

The usual method for dealing with what you don't know is to consult an expert. People do that with physicians, auto mechanics, and plumbers all the time. It's mostly okay. But there are cases where it isn't. Climate is one of those. Oh yes, I know we can all understand droughts, heat waves, snow, floods, storms, and the wrath of the gods. But how do you know whether any of these phenomena are related to climate change? This isn't a plumbing issue. Even if you don't do your own plumbing, you know when your toilet does not work and when it's fixed afterward. That's crucial. After farming out everything else to experts, knowing whether you got what you want or need is your last link to your own problem. It keeps you in charge.

However, if you farm out something like climate to experts, where you have no expertise yourself outside of some prejudice and folklore, you are like a child, precisely because you cannot say, on your own, whether or not there is a problem, let alone whether it has been fixed or not. So, why not just put the experts in charge then? You would only vote according to what they tell you anyway. Isn't that the concept? Think about that. If you don't step up to learn new things when needed, what starts with plumbing flushes away democracy.

But, cluelessness notwithstanding, children and cavemen insist on decision-making anyway. There are a number of popular methods. The most primal of these is to decide whether you trust some particular expert. Where would you begin? Maybe you can discover whether the expert had any shady real estate dealings. Were there extramarital affairs, prior convictions, drunk driving charges? How about cheating on taxes? Exposing moral turpitude as a way to make up minds about climate experts may sound a bit hyperbolic, but it isn't. It's the first approach to the climate problem. Aficionados decide on physics through moral turpitude. Reality check: that should sound bonkers.

The moral turpitude of choice is fraud. Allegations are made that false opinions are bought and paid for from experts by forces that aim to deceive. If identified, the next step is to just ignore the offending experts, after some primal Paleolithic vilification. Simple enough. But how can you find fraudulent experts out? You could turn to an 'expert' at investigating the experts. But this is just another expert problem. Do you trust the experts on experts? Is this a concern? Well, yes. There is a rich industry of activists anxious to share their 'expertise' in implicating putative climate experts in evil deeds. But thoughtful people soon catch on that something is amiss.

Many of their claims are outrageous smears that are probably actionable. Nevertheless, celebrities, government officials and even heads of state deal freely in them. It's fashion, like narrow lapels and short skirts. There are any number of libelous websites that claim to expose climate experts who do not practice 'right thinking.' Why is this injustice spree allowed to stand? Ironically, despite allegations of being on the take, many of the libelous sites could probably be closed down if the experts in question actually had money for libel actions.

Real climate experts are real scientists. They don't test the morality of experts to decide what's true in nature, because they don't believe any experts. That's science. They don't worry about deception, because we humans do an excellent job at getting things wrong even without lies! For real scientists, the issue is finding out the falsehoods irrespective of motives. Test things, not persons. That's science at its best. It follows that real scientists find the moral turpitude test, and the accompanying primal politics, bewildering. It's a farce to them. So let's set moral turpitude aside.

Another popular tactic for asserting control, when you know nothing, is to test qualifications: the competency method. Maybe you can decide on which expert to believe based on whether the expert is qualified. I once heard from a journalist who was compiling a list of those qualified to speak publicly about climate. His dream, simply put, was that qualified experts speak and everyone else shuts up. He approached me to determine whether I should be on his list.

I asked him what made him qualified to decide who was qualified. Silence. It borders on a self-referential paradox. Who is qualified to choose who is qualified to compile such a list? It wasn't even necessary to point out that self-appointed qualification deciders might just rule out experts holding views they don't like. So much for the competency method.

It would be simpler if all experts were to say the same thing. This is easily achieved, if you are up for some mathematical madness. Simply define experts as those holding a particular prescribed position. All experts then agree by definition. Other views, equally by definition, are not from experts. It's simple. Forget about distinguishing between experts; they're all alike, by definition. Your contribution to important discussions need only be to reply to doubters with, 'The experts all agree.' It's liberating. Who cares who agreed or what they might have agreed to? Try it: ask people who and what. It's no wonder that climate journalists, who have never heard of the equations that govern the motions of air and water, could survive despite a quarter century of boiling climate fervor.

But, despite the madness of it, if you don't actually know anything, how can this even be achieved? Any prescription for what's 'correct' requires at least some knowledge to set the definition. Maybe you could farm that out too. But you do need to be reassured that some canonical position has actually been set by someone, even if you avoid learning what it is. Moreover, you still need to know specific tasks that you are called on to do by the inscrutable, duly defined experts. While the Paleolithics might not be able to help you with microwave ovens, they would recognise at once who you need. You need a shaman to intercede for you with the gods. The experts recede into the background as unknowable entities of power (the gods). They're interpreted for you by some bold, charismatic personality (the shaman). Al Gore is a shaman.

So in the end, despite our democracy, modernity, sophistication, and technology, we return to the old stone age, because it all boils down to what the shaman tells us the gods say, while the supplicants chant, 'Club. Club. Club … ' And if you do not agree that this is 'science,' then you believe in smoking and that the earth is flat, or something like that. But the modern supplicants actually chant something else equally devoid of scientific content: 'climate change is real; the science is clear; the scientists all agree.' I suggest you look off into the distance, hold your arms wide, and repeat it full voiced, three times, to get the full inspirational effect. I've tried it. It's a compelling ritual. Perhaps it could be put to music.

Enablers and oracles

These methods aren't for everyone. But there's a fourth, very hip, alternative. It was once explained to me by a certain sociologist. She told me that it did not matter that she knew nothing about science. She knew what to do anyway. Her explanation was so smooth it was like a song:

Don't know much about climatology;
Don't know much about astronomy;
Don't know much about a physics book;
Don't know much about the math I took;
But I do know what we should do
And I know that if you do too
What a wonderful world this would be.

But no matter how mellifluous the explanation, reasoning matters. If we don't act, she reasoned, things may or may not turn out bad, but if we do act then things will be fine. Therefore we must act. Her thinking is sometimes described as the 'precautionary principle.' The precautionary principle originates from medicine, where it's more of a do-no-harm kind of thing. Inaction, instead of action, is the prudent course there. The climate version is upside down. So instead we'll call it the 'Wonderful World Method' (WWM) for ditching science.

It takes talent to make the science in a scientific issue disappear, which WWM does. But where did it go? If the WWM cases are put into a table, listing 'action' and 'inaction' against 'bad' and 'fine', there are four possibilities, not three. In the WWM, the 'action and bad' case is missing. The missing case seems like a stray thread, but if you pull on it everything unravels. The scientific questions didn't disappear; they were just embroidered over.

There are surely more methods to avoid learning the science needed. As any parent knows, there are boundless excuses to avoid your vegetables, and there are also always enablers who feel it's cruel to make the kiddies eat their vegetables. The climate policy process is a no-vegetable enabler's paradise. The process envisions policymakers standing before the single closed door of an otherwise sealed room.

A note, written in mystic runes, is slipped under the door. Imagine visiting an oracle in ancient Greece. Policymakers leave with the word of the gods. It's so simple: scientists figure out science; policymakers listen to what they have to say; policymakers make policy that doesn't contradict the laws of nature. Then we all ride off into the sunset basking in a warm utopian glow. Problem solved. Vegetables successfully uneaten.

This process embraces all four methods for ducking science. The unseen within the sealed room are the gods. The policymakers visiting the oracle are the shamen. Who is selected to be in the sealed room? Well, only competent experts of unquestioned morality (as long as they're in the room) are selected. Who selects them? Only the shamen know.

But where's WWM here? This is the really interesting part. No one actually cares about what the oracle wrote, let alone why. In reality, of course, the note is not a piece of paper, but rather a huge tome: the scientific report of the UN's climate panel. Not only do few ever read it, but policymakers actually doctor it after they receive it. That is, as the policymakers walk away from the door with the note, they scratch out some lines, and reword parts, while inserting little somethings here and there. That way the note will agree with what they think the oracle ought to have said.

What? You didn't know? It's no secret. It's standard operating procedure for the UN panel. What difference does it make! WWM makes the scientific issues irrelevant anyway. The frank purpose of the UN climate process, according to its founding head, is to 'orchestrate' scientific opinion. The current head is less frank, uttering things like 'Climate change is real' instead. I don't think he sings it to music though.

On being smarter than a screwdriver

People are much smarter than they are given credit for, when given half a chance, a bit of encouragement, and not trumped by fashion. If everyone involved put as much effort into learning and explaining over the last 25 years as they put into avoiding, enabling, and group-think, we would not be living the life Kafka. But no matter how Kafkaesque our approach to this problem is, an inexplicable toehold remains when the nonsense is alleged to be backed by computers.

Computer cachet is a pernicious part of the superstitious cloud of confusion stopping us from thinking. Computers aren't oracles either. They are just tools, like a screwdriver. The first rule when using a screwdriver is to be smarter than the screwdriver. You can hurt yourself if you aren't. It's the same for computers. The question is sometimes put to me as to whether computer climate models are ready for policymaking. Climate models are the best we have, but they are far from good enough. Even many experts are unaware of the extent of their limitations.

Computers can only hold a finite number of numbers. Computer scientists call that a 'finite representation'. Because of that you can get garbage out even when you don't put garbage in. It alters arithmetic; it alters the equations themselves, and it means important physics has to be faked because modern computers are far too slow and their representations are far too small for climate. No climate model fully employs the known physics. They are empirical. But climate forecasting is not an empirical problem. If one has a computer large enough, it is easy to estimate how long a typical modern computer would take to do one 10-year forecast without some of this fake (empirical) physics. With a Kolmogorov microscale of about a millimetre for air, one gets numbers like 100,000,000,000,000,000,000 years. That is longer than it took for Douglas Adams' famous fictional computer, Deep Thought, to answer the cosmic question.

I could elaborate, but in a time of neo-shamanism, moral turpitude experts, and WWM, does it matter? The climate fervor was never about 'science'. That word is just a gimmick and a weapon. Policymakers can get little from computer climate models if they fail to grasp their deep, unredeemable limitations, while being distracted by para-scientific agendas. It's no good wondering whether climate models are ready for policy, when most policymakers are not ready for models. They are not smarter than this screwdriver.

18   The scientists and the apocalypse

Bernie Lewin

The meeting of the United Nations Intergovernmental Panel on Climate Change (IPCC) in Sundsvall, Sweden, August 1990, witnessed a Third World revolt that was premeditated and forewarned. It had already begun in the previous working group meetings set to develop international policy responses to the climate crisis. But only in Sundsvall, under the leadership of Brazil, did it succeed in smashing this carefully conceived science-to-policy process at its very nexus. Within months the revolution was complete.

At the United Nations General Assembly that December, the climate treaty process was taken from the IPCC and its UN parent bodies—the Environment Program (UNEP) and the World Meteorological Organisation (WMO). Instead, a new negotiating committee would report directly to the General Assembly, where the poor countries commanded an overwhelming majority. The IPCC, dominated by scientists from rich countries, was directed to serve this new committee in the interim, until a subsidiary body for technical advice could be established. As for the two peak science-policy organisations who first conceived the IPCC, by winter 1991 they were out in the cold.

This banishment from the treaty process was particularly shocking for UNEP. In the afterglow of its success with the ozone treaty, it was coming up to the twentieth anniversary of its inception at the 1972 UN Stockholm conference where global environmentalism was born. Riding a new wave of environmental consciousness, another grand conference was in the planning to mark the anniversary. The Rio 'Earth Summit' of 1992 would be the biggest UN talkfest to date, with its policy centre piece, the Framework Convention on Climate Change (FCCC). But few would ever guess just how much this Convention was a political triumph not for UNEP but for the conference hosts, Brazil. Its success would set in train the role of the poor countries in the climate treaty negotiations where the talks stalled and stalled again with their repeated attempts to use the pretext of warming mitigation to increase the flow of aid.

In *The Age of Global Warming* Rupert Darwall details how global environmentalism concentrated itself onto the global warming scare.[1] Here we take up with a group of activist climate scientists, tracing how they entered this political game, how the greater politics of the UN quickly overwhelmed and corrupted their science, and, finally, how the academies of science were soon dragged down with them.

Science-for-policy

Behind the establishment of the IPCC was a conscious effort by some scientists to get the science-for-policy mechanism right. Pitfalls at the policy interface were all too familiar to those involved. On the one hand there were always scientists with extreme views capturing media attention and generating an inflated sense of alarm. When the intergovernmental assessment panel was first formally proposed at the tenth World Meteorological Congress in May 1987, the ice age scare of the 1970s was well remembered. Asked to assess the science behind this scare, national scientific academies delivered sobering reports and the WMO itself made some effort to quell alarm.[2] But then there were those also concerned that well-founded alarm might go unheeded, especially when a global problem requires a cooperative response. The IPCC was to provide the balance.

The basic design was for two distinct tiers: first the scientific assessment and then the science-policy interface. In the first tier were elected experts from the relevant fields assessing the current state of the science. They drafted a report, circulated it for peer review and then redrafted it in response to that review. The second tier involved science-literate government delegates agreeing on a plain language summary of the report. This 'Policymakers Summary' would be, at the same time, grounded in science, policy relevant, yet politically neutral and agreed by all.

The workload of the assessment was divided between three working groups: Working Group I to assess the scientific basis of warming concerns; Working Group II, to assess impacts of this warming; and Working Group III, to consider response strategies, whether mitigation or adaptation. When the expert authors of each group submitted their completed report they would also provide a drafted summary. This would then be finalised by consensus at a meeting of the delegations. All this was to take place before the three parts of the report arrived at a full session of the IPCC. There, a further summary of these summaries would be negotiated to consensus, thereby producing a peak document for delivery into the policy debate. Indeed, as the process came to completion, just such a 'synthesis report' was drafted by the renowned atmospheric scientist chairing the IPCC, Bert Bolin.

At only the fourth full session of the IPCC it was even partially approved before the Brazilian-led revolt left it in tatters on the floor of that conference hall in Sundsvall. But that is getting ahead of ourselves.

Greenhouse postponed

The IPCC Working Group I had the easiest task as they trod a familiar path of US government assessments back through the 1980s.[3] But it was another international assessment conducted by another international science body that served as the principal model for their report.

The International Council of Scientific Unions (ICSU) is constituted of representatives from most of the world's national science academies and the various international scientific unions. It had long coordinated international geophysical research before establishing a Scientific Committee on Problems of the Environment (SCOPE) to report to the Stockholm conference. Beyond Stockholm, the SCOPE committee continued to commission environmental assessments, including one on the greenhouse effect, *SCOPE 29*, completed in 1985.[4] Not only would this be the forerunner to the IPCC scientific assessment, it was also the basis for a conference in the Austrian village of Villach that is often hailed as the birthplace of the climate treaty movement.

Convened in autumn 1985 by the ICSU, along with UNEP and WMO, this conference generated momentum for urgent global action. The UNEP executive director, Mustafa Tolba, made a particularly strong case to the 89 invited experts for the commencement of another treaty process like he was then facilitating to protect the ozone layer.[5] The final agreed statement concluded that greenhouse warming was 'expected,' that it 'appear[ed] inevitable,' and that the prospect of catastrophic warming—with the doubling of greenhouse gas concentrations as early as 2030—required urgent mitigating action.[6] Yet the SCOPE report itself was of quite a different tenor. Sure, the computer modelling predicts the usual range of warming for an equivalent doubling of CO_2 (1.5-5.5°C), but this remains nothing more than theoretical speculation until it is validated empirically. The section where the progress of this validation is assessed offers the least cause for alarm.

During the 1980s, there was much apprehension in government reviews and elsewhere about claims based on what the Villach statement calls 'advanced experiments with general circulation models.' But even the modelling advocates recognised that policy commitment would be unlikely without empirical evidence that, firstly, variations in atmospheric CO_2 concentrations had a significant impact on global climate, and, secondly, that the contribution of industrial emissions was already having such an impact. The detection of even the slightest suggestion of an anthropogenic influence might have been sufficient grounds for policy action but it would surely be the minimum grounds.[7] All the government-run assessments agreed that the theoretical cause-effect relationship between emissions and warming had not been established, and this international assessment for ICSU presented no exception.

The 'Empirical Climate Studies' section of the SCOPE report was compiled at the Climatic Research Unit, University of Norwich, under Tom Wigley, a leader of 'detection' research throughout the 1980s. Wigley found not only that the influence of emissions remained undetected, but also that evidence was insufficient for natural CO_2 variations causing changes in the past. When, in 1989, Wigley was asked to coordinate the drafting of the 'detection' chapter, Chapter 8, of the IPCC report, no new evidence persuaded him from these conclusions. The preceding chapter in the IPCC report also discussed persuasive evidence in the climate record running contrary to the hypothesis that CO_2 was a significant driver of change.[8]

So, with the IPCC Working Group I report finalised in 1990, we had a second international assessment in five years coming to the same conclusions on the empirical science that we found in all the various governmental assessments during the preceding decade. In between, James Hansen of NASA made front page news by calling for climate action after giving testimony to US Congress of 99 per cent certainty that man-made climate change is happening now. But with his performance followed by a wave of remonstration among his peers, the IPCC report only serves to confirm it as the grandstanding of an extremist.[9]

Nevertheless, Hansen's call-to-action only reiterated the plea from Villach. And when *SCOPE 29* was published in 1986, the Villach statement had been placed at its front, as though an executive summary of the non-prescriptive, equivocating underlying report.

With the IPCC, it was a different story. The summary at its front, as agreed by all the governments, was notable for its fidelity with the underlying report. Sure, some speculation was puffed-up to appear more solid, but mostly it remained true to a report that hardly constitutes the scientific basis for urgent and drastic action.[10] In other words, and to the credit of its designers, the science-to-policy process had worked.

Consider also how, in the review process, Wigley was asked to consider when the human influence will start to become apparent. In 1981 he had suggested this likely around 2000, but by 1990 he was not so confident.[11] Only when the half a degree of warming seen in the first half of the twentieth century is again repeated, concluded Wigley, will we then be likely to determine just how much of it is human-induced.[12] The IPCC also moderated the expected rate of greenhouse gas build-up, and so, according to the worst-case modelling, this extra warming was not expected for at least a decade, but more likely not for many decades. In other words, not only had detection been postponed, but (as Brian O'Brien was quick to declare) so too had the whole emergency.[13]

Alas, by that stage no one was listening. Three years after the IPCC process was first conceived, the working group set to establish the scientific basis for action found itself the calm centre of a maelstrom of climate enthusiasm brewing all around. As it would eventually break the science-to-policy process, let's go back and consider that brewing storm.

Warming enthusiasm

The IPCC was conceived before the frenzy for action whipped up during the hot dry North American summer of 1988, but it was then born into the storm of enthusiasm that ensued. In the first place there was the grandstanding of extremists like Hansen. On top of that came minor state leaders, like Gro Brundtland (Norway) and Brian Mulroney (Canada), channelling the enthusiasm for sustainable development—something of a campaign yet to find its cause. Then, topping it all, came Margaret Thatcher propelling climate action up the agenda of the Group of 7 talks.[14]

And yet, after all the excitement of that summer, the first session of the IPCC in the autumn remained a relatively low-key affair. Only 30 delegations arrived, and these mostly from northern countries already active in the research.[15] What really drove political interest came only weeks later with the first resolution committing the UN to protect the global climate. This resolution did more than endorse the new UNEP-WMO panel. The IPCC was also asked to make recommendations on the use of 'relevant existing international legal instruments' and to explore 'elements for inclusion in a possible future international convention on climate' (Article 10).[16]

And so begins 1989. The year the Cold War ended was extraordinary for the intensification of interest in climate action at the intergovernmental level. While Wigley and all the other authors were busy taking their chapters through the review process, numerous ministerial conferences convened around the world to address the climate emergency. Their concluding statements competed for the louder alarm while participants queued to pledge themselves for saving the planet. Already in January, a harbinger of the pending onslaught appeared before the IPCC. The US Secretary of State for the new Bush administration, James Baker, chose for his first public engagement to open a session of their Working Group III with a call for political action.[17] From the beginning, political interest was concentrated on this 'policy' group—as Working Group III was often called—and the USA had already won its chair. With this new UN resolution pushing the IPCC further into the domain of policy development, the intensifying political interest concentrated overwhelming pressure on this group.

However unwelcome this premature political excitement might be for Bolin as he tried to deliver for it some *scientific* grounding, the effect was that Tolba's vision of a climate treaty process now realised sufficient *political* ground.[18] And so Tolba formally submitted a plan for UNEP to work with WMO and the IPCC towards delivering a framework convention.

The UNEP plan was approved by the General Assembly in December

1989,[19] and from early 1990, the detail was widely known and generally agreed: in August the IPCC would approve its 'interim' report in Sundsvall before presenting it to the second World Climate Conference in November and to the General Assembly in December; then, early in 1991, its parent bodies would open negotiations to develop a framework convention so that an agreed text could be ready for signing at UNEP's anniversary 'Earth Summit' in 1992.

In Washington, February 1990, attendance at the third full session of the IPCC had swollen to 260 delegates from 62 countries. President Bush opened proceedings with an enthusiastic speech in which he demonstrated his commitment to the UNEP plan by repeating an offer to host the first session of Convention negotiations.[20] Throughout the year, WMO and UNEP busily prepared for these negotiations in various ways, even by convening an *ad hoc* planning group of government representatives.[21] They met only once, just after Sundsvall, but for all appearances at least, the UNEP plan was on track right up to the World Climate Conference. There, Thatcher took the completed and approved IPCC report as 'our signpost', directing UNEP and WMO, which were 'the principal vehicles' taking us to 'our destination', a climate treaty.[22]

As we know, the poor countries alliance made sure that these vehicles never hit that road. *What was their beef?*

The aspirations of the poor

In the early days of the UN, its impoverished members were in fragmented minorities, but, as decolonisation progressed, a majority emerged and began to assert itself in the trade talks as the Group of 77. On the environmental front, even before Stockholm, it became clear that coordinated global action would only be possible if made attractive to governments of the decolonised world. This generally meant some commitment to a flow of resources from rich to poor, but with some sort of legitimation looking better than a ransom payment. And that still left the problem of economic growth. The conflict between environment goals and the development goals of the so-called 'developing' world had to be resolved. And so it was: in the doctrine of sustainable development.[23]

The sustainable development philosophy promoted famously in the 'Brundtland Report' (1987) would be epitomised in the 'Rio Goals' and 'Agenda 21' of the Earth Summit, but by then it was already well established in the climate talks, as is evidenced by the Framework Convention. This document defined the differential commitments of its signatories across the developed/developing divide and called for the one side to help the other towards their common future through transfer of money and technology. 'Technological transfer,' as a condition of agreement, was not new even with the climate negotiation. Throughout the 1980s the UN Law of the Sea talks had been locked down by such demands. If only under a new banner the climate talks were similarly stalled with the bigger players, including Brazil, India and China, pitched against the USA. Not only should legally binding emissions targets be limited to those who have caused the problem, but already before Sundsvall in the policy discussion of Working Group III there were demands that the post-industrial rich sponsor a more sustainable industrial development across the impoverished regions of the world.[24] The IPCC process had begun with the scientific leadership ready to address the sensitivities of development politics. Their first problem was to get delegations to show up. A trust was soon established to fund attendance and a special committee convened to support the active participation of those struggling even to find science-literate delegates. It was only with this success that the trouble began.

The revolt

August 1990, and the conflict in the policy group arrived unresolved in Sundsvall with Brazil proclaiming its intentions before the fourth full session of the IPCC even began. Disrupting the approval process with novel amendments, they blocked consensus on the synthesis report. As the dishevelled proceedings dragged on into the evening of the final day, it seemed a sabotage attempt had succeeded. But in the desperate small hours of the night, Bolin managed to salvage a summary of sorts. This was not the one he had drafted, but mostly a patchwork of extracts from the already approved working group summaries. There were also two novel insertions: one, a disclaimer that the report does not reflect the positions of all participating governments; the other, a declaration that rapid technological transfer is urgently required.[25]

An embarrassment to the process, this synthesis was never widely distributed and only ever published two years later behind another report. Never again would the IPCC delve into the fraught domain of policy proposals, henceforth restricting itself to 'policy-relevant' advice. But before it was able to produce another full report, the struggle for control of the policy agenda threatened its very survival.

The December 1990 resolution to replace UNEP management of the treaty process with an independent International Negotiating Committee (INC) did still name the IPCC as a source of technical advice.[26] But, in the drafting of the FCCC by this committee, this emerged as only an 'interim' arrangement, until the 'conference of parties to the convention' (COP) could establish its own Subsidiary Body of Scientific and Technical Advice (SBSTA). The purpose of this body would be to provide 'assessments of the state of scientific knowledge.' SBSTA would 'prepare scientific assessments' and it would respond to any technological questions of the parties. The wording could not be clearer: at the first Conference of Parties meeting (COP1), SBSTA would replace the IPCC.[27]

Alive and walking Frankenstein

As its future was thrown into doubt, the IPCC had another problem, which was in the doubtful advice it continued to provide. Under the interim arrangement, in the lead-up to the Earth Summit, and with its policy group languishing in stunned suspension, the IPCC continued to assert its independence by preparing supplements to the assessments of the other two working groups. Added to the top of the Working Group I supplement was an updated summary of the science. Its six cautious 'major conclusions' only showed how little had changed, with the last repeating the prediction that detection is not likely 'for a decade or more'.[28] While the IPCC confirms that it was stuck on basic questions, the INC was writing its equivocations into obsolescence.

The Framework Convention on Climate Change (FCCC) opened with a statement that emissions 'will result in' global warming. There seemed to be no need for Working Group I to continue its search for empirical confirmation. The FCCC then redefined 'climate change' as only that resulting from human activity. In one stroke the prior work of the IPCC was reframed to its needs. Not surprisingly, whenever the INC did solicit further advice, it was never to confirm the science, but only ever to inform the development of climate treaty protocols.

One such question, perhaps the most basic of all, caused no end of friction between these two UN organisations. The objective of the FCCC was set to 'stabilise greenhouse gas concentrations in the atmosphere at a level that would prevent dangerous anthropogenic interference with the climate system', and so the IPCC was asked to provide the level at which such concentrations would become dangerous. But as discussions began, with special workshops convened, Bolin became increasingly preoccupied with the uncertainties engulfing every aspect of this question.

The INC kept pressing for an answer while Bolin shifted towards the view that this was not a question for the IPCC because it was in the realm of policy. While others in the IPCC leadership disagreed, he slipped into further doubt, wondering whether anyone in any sphere could provide a meaningful answer. With so many uncertainties in the relations between chaotic systems, both natural and social, it would be impossible to make a call on where danger lay.[29]

Nevertheless, the IPCC pressed ahead with plans for a second full and independent assessment, scheduled for release in 1995. This might have come in before the interim arrangement ended, with COP1 required within a year after reaching a quota of Convention ratifications. When this was attained earlier than many expected, in March 1994, the timetable for the second assessment review process had already pushed out to the end of 1995, way beyond COP1, scheduled for that March.

Early friction with the INC did not amount to much until 1993 when Raúl Estrada-Ouyela, an Argentinian diplomat, was elected to the chair. In preparing for COP1, at COP1, and beyond, Estrada made it known that Bolin and his IPCC were an annoyance that the treaty process would be better without. After the COP1 date was fixed, Estrada repeatedly asked Bolin to fast-track the new assessment. It was when Bolin deferred a long-planned joint meeting on the 'dangerous levels' question that Estrada wondered aloud whether the scientists were 'suffering from a Dr Frankenstein syndrome.' Their work had created the Convention, yet now that it is 'alive and walking, and deciding things, the scientists have reacted against its demands.'[30]

Were they afraid that their creation is now dangerously out of control? Estrada teased the IPCC with this analogy while visiting the UK, and so *New Scientist* asked the British chairman of Working Group I, John Houghton, to respond. Houghton defended their tardiness just as Bolin had: the rigorous peer-review process needed time to run its course. But anyway, he could see no rush, for the new report would contain little that was new. 'The past four years' work has underlined and confirmed most of what we said in 1990.'[31]

That was in June 1994. Early in 1995 COP1 produced the 'Berlin Mandate' and SBSTA. Meanwhile, the IPCC, now on tenuous authority, continued with its assessment.[32] In the summer, Working Group I lead authors met for the last time in Asheville, North Carolina, to finalise the chapters and draft the summary. When these were circulated ahead of the approval plenary scheduled for November in Madrid, it was clear that Houghton was right—not much had changed. Not much except something of a retreat on the question of detection. New research by the lead authors of the detection chapter raised new concerns about determining natural variability from the patchy and uncertain climate record. With inadequate knowledge of natural variability, there was no 'yardstick' against which to measure the human influence.[33] The detection chapter, Chapter 8, closed off with a deeply sceptical conclusion. The retreat to scepticism is complete in the response to the question of when detection might be achieved: 'We don't know.'[34]

The drafted Policymakers Summary was not so sceptical and cheer-leading journalists let everyone know by leaking a weak detection claim.[35] With legally binding emissions protocols now a real prospect, there were business interests keen to see it moderated. Everyone was ready for a real showdown in Madrid and this time the business lobby was well prepared.

Scientists against science

Especially following the Rio Summit, non-government organisations of all shades had been encouraged to participate in the review process. Opposite Greenpeace, a business coalition that they called 'The Carbon Club' had already made a strong case for weakening the summary with a raft of suggested changes built on the very wording of the underlying chapters.[36] At Madrid, they continued the campaign in collaboration with the Saudis. But not long after the meeting opened they were thrown completely off guard.

From the Chair, Houghton told the assembled delegates that there had been new positive developments in the science of detection which the coordinating lead author of Chapter 8, Ben Santer, would explain in an extended presentation. This concluded with Santer announcing that the chapter, as written, was out-of-date. Then Houghton established a side-group to draft an update of the working group's findings.

When this procedure was challenged, there was a ruling from the Chair. This was challenged again and Bolin lent his support. A Saudi delegate protested that during his six-year involvement with the IPCC the underlying report had always been the authority for the summary. But now it appeared that the rules had changed. Indeed, they had.

In the short history of the IPCC there had always been pressure to bring the report into alignment with some agreement on the floor. But this was always met with strong resistance from the Chair. Of course, there was Sundsvall last time around. But, even as the Saudis protested in Madrid, there remained an unresolved dispute in the reformed Working Group III.

For the Second Assessment, this cursed working group was asked to delve into the neglected economic dimensions. Troubled from the beginning, the approval plenary broke down, was reconvened, but the crisis continued over calls to change a chapter. The problem was in the estimations of climate change damages, where the value of a life among the rich was taken as ten times greater than among the poor—where most deaths were expected. Moreover, devaluing the poor so reduced the damages that this was seen to justify the air-conditioned rich doing nothing about their continuing impacts on the rest. This 'price of life' controversy broke out at COP1, where it exacerbated the rich/poor polarisation and embarrassed the IPCC just as its interim advisory role expired. And it would continue even after the Working Group I plenary in Madrid, with calls to break the deadlock by chapter changes only met by the authors' refusals upheld by the Chair.[37] At Madrid it was a different story. There, the call to break with the science-to-policy process came from the Chair.

The new developments Santer presented to delegates in Madrid were from his own recent and unpublished work. While assessing their implications for the problem of detection just before the lead author meeting in Asheville, Santer said: 'I don't think this is evidence that we've solved the problem; far from it.'[38] At Asheville he was not so reticent and he used a special presentation of his findings to push for a positive detection 'bottom line'. This met with strong resistance across the four-day conference and beyond. As a result, the draft summary finalised after Asheville contained a weak detection claim glaringly inconsistent with other passages and with Chapter 8's sceptical conclusion.[39] In Madrid, despite further protestations through the mouthpiece of the Saudis, this inconsistency was resolved in favour of detection. After the meeting, the chapter's concluding summary was removed along with the 'We don't know,' and other sceptical statements; thus smoothing the way for the bottom line detection claim already agreed in Madrid: 'the balance of evidence suggests a discernable human influence on global climate.'

A discernable human influence

However so weak this detection claim, it had enormous impact as the slogan for the treaty talks to take the next step. At COP2, when the USA finally gave way with support for binding emissions protocols, it was presented as the scientific basis for their decision. Had it not got up in Madrid, 'governments would have faltered on taking urgent action … such as signing in 1997 the Kyoto Protocol'—so Houghton later reflected in the prestigious scientific journal, *Nature*, under the banner: 'meetings that changed the world'.[40] Perhaps. Perhaps only as a convenient rhetorical authority, but its timely provision made a world of difference for the IPCC. By political and media acclaim, detection launched the IPCC back onto centre stage, where its role as the permanent scientific advisor to the treaty process passed beyond question. The pretender, SBSTA, was soon reduced to the mere conduit of its advice. But this marvellous renaissance had a dark side.

The Carbon Club were none too happy with the way they had been outmanoeuvred. Retaliations began as soon as the Club got hold of the new version of Chapter 8; the scandal breaking just before COP2.[41]

Their case was devastating and difficult to refute, yet passes unsupported within the scientific establishment. All support was for Houghton to dismiss the ongoing protests as merely the wailing of the unrepentant global polluters vanquished by the scientists in Madrid—it was as though the saints of the apocalypse had won a mighty battle to secure our common future.[42] Such heroic resonances might have saved face, but redemption's great price would be paid and paid again and across the institutions of science; an exceptionally explicit expression of which is a volte-face caught freeze-framed in *Nature*.

During 1995, *Nature* ran numerous embarrassing news stories on the various scandals emerging during the Second Assessment, including one lead author's refusal to release to a reviewer the modelling data behind a key (but dubious) graph.[43] Two scathing editorials attacked the IPCC, one calling for the first working group to be reined in to 'a more judicial course' while suspending the others.[44] When the Chapter 8 controversy broke, the barrage continued with an editorial in support of the complainant and doubtful about the defence. And yet this time the critique was couched in restraint. Indeed, the main thrust of the editorial was a call for restraint—its headline imploring 'Climate debate must not overheat.' Why? Because these charges against the IPCC 'should not be allowed to undermine efforts to win political support for abatement strategies.'[45]

Those scientists failing to heed this advice were marginalised and ostracised as traitors to science, unsupported by their academies. Scepticism became intolerable, silence golden. After Madrid, during the late 1990s, energy and chemical companies ran from the Carbon Club to a fully green image. Emboldened, scientific institutions saw fit to break their silence, standing up for virtuous science and the common good. Most significant historically was when the president of the oldest state-sponsored empirical science academy, the Royal Society, launched it into the new millennium with a declaration for the cause, thus violating their ancient policy 'never to give their opinion as a body.'[46] By 2007, the American Physics Society was proclaiming global warming a truth 'incontrovertible'.[47] Given that every physicist knows that in empirical science there is nothing beyond challenge, their declaration was surely the clearest message of all that what was being defended here was not science but dogma.

And so we have it. Starting with a handful of scientists in obscure fields sucked up into the save-the-world politics of UNEP, we have now arrived, after only a few short decades, with corruption spread across our great institutions of science. Whether knowingly or not, and whatever their motivation, these scientists opened a gap for huge political forces to overwhelm their principles and processes, and to empower those among them willing to participate in the corruption.

19   The scientific method (and other heresies)

Stewart W. Franks

Perhaps the most frustrating aspect of the science of climate change is the lack of any real substance in attempts to justify the hypothesis. Often the occurrence of a drought or flood is sufficient to generate a whole range of expert speculation from those that should know better. Often the claims are couched with the disclaimer, 'Of course, no single event is attributable to carbon dioxide and climate change however this is exactly the sort of event we would expect to see.' Such statements are meaningless, as these are also the sort of event that we expect to see irrespective of anthropogenic climate change. What is most dismaying is that the worst examples of speculative claims often come from the scientists themselves.

Commentators from the Bureau of Meteorology and the Commonwealth Scientific and Industrial Research Organisation (CSIRO) are amongst the worst for making statements that are simply incorrect.

One stated, 'Of course, the drought has not been helped by rising temperatures, which have increased losses through evaporation,' and, 'It is very difficult to make a case that this is just simply a run of bad luck driven by a natural cycle and that a return to more normal rainfall is inevitable, as some would hope.'[1] In an interview with the *Sydney Morning Herald*, another commentator mused that 'Perhaps we should call it our new climate.'[2] A similar line was adopted by another climatologist, 'In the minds of a lot of people, the rainfall we had in the 1950s, 1960s and 1970s was a benchmark … But we are just not going to have that sort of good rain again as long as the system is warming up.'[3]

Such claims are not limited to off-the-cuff speculation in the media—many incorrect claims surface in the scientific literature. One example was the flurry of activity attempting to link the recent Murray-Darling Basin drought and anthropogenic climate change.

The Murray-Darling Basin drought was initiated by the 2002-3 El Niño. Immediately following this, a report under the auspices of the WWF-Australia noted that whilst rainfall had been low, the air temperatures had been particularly elevated.[4] This led the authors to claim that:

> The higher temperatures caused a marked increase in evaporation rates, which sped up the loss of soil moisture and the drying of vegetation and watercourses. This is the first drought in Australia where the impact of human-induced global warming can be clearly observed.[5]

While this may sound intuitively correct, it is wrong. It completely ignores the known science of evapotranspiration and boundary layer meteorology. That is, when soil contains high moisture content, much of the sun's energy is used in evaporation and consequently there is limited heating of the surface. However, during drought, soil moisture content is low and consequently nearly all of the incoming radiation is converted into heating the surface. The result is that air temperatures rise significantly.[6]

Others went further. CSIRO researchers published a number of studies in leading international journals which wholeheartedly adopted the flawed physics to make some spectacular claims. The researchers claimed that a one degree increase in temperature causes a reduction in annual water flows of fifteen per cent.[7] The series of reports and journal articles confused cause and effect with regard to the fundamental basics of evaporation and boundary layer meteorology. Reduced evaporation causes higher air temperatures. It is drought which is naturally associated with lower evapotranspiration which leads to higher temperatures. Above all, there was no expertise evident in these claims.

Such misunderstanding of basic physics is not confined to Australia's lead scientific research organisations. US and European studies of trends in drought have also confused the role of air temperature and evaporation.[8] These studies employed the Palmer Drought Severity Index (PDSI) which approximates evapotranspiration through a simplistic model that utilises temperature as a proxy for the atmospheric demand for moisture. Given an increasing trend in air temperature, the studies would demonstrate an increasing trend in evaporative losses. Again, such claims were readily reported in the media. Eventually, in 2012 Sheffield et al. identified the error in such a simplistic approach, stating:

> The simplicity of the PDSI, which is calculated from a simple water-balance model forced by monthly precipitation and temperature data, makes it an attractive tool in large-scale drought assessments, but may give biased results in the context of climate change. Here we show that the previously reported increase in global drought is overestimated because the PDSI uses a simplified model of potential evaporation that responds only to changes in temperature and thus responds incorrectly to global warming in recent decades. More realistic calculations, based on the underlying physical principles that take into account changes in available energy, humidity and wind speed, suggest that there has been little change in drought over the past 60 years.[9]

Unfortunately, the error received little if any attention in the media and the public, and the misconception of 'observed' increasing drought due to anthropogenic climate change continues to be repeated in public and scientific discourse.

As with all droughts, the Murray-Darling Basin drought came to an end—and as so often in Australia, it took a flood to break the drought. Despite the speculation noted earlier that 'In the minds of a lot of people, the rainfall we had in the 1950s, 1960s and 1970s was a benchmark … But we are just not going to have that sort of good rain again as long as the system is warming up', eastern Australia experienced rainfall all too reminiscent of the 1950s, 1960s and 1970s. In late 2010, La Niña returned (after a nine-year absence) and with it we saw flooding across eastern Australia, the likes of which we had not seen since 1974 (also a La Niña year). Whilst the floods and cyclone were devastating in their impact, again there was nothing to suggest that their occurrence was anything but normal given the known climatology of eastern Australia. But once again, the experts lined up to proclaim that, no doubt, climate change was involved. It seems that any climate extreme represents the impact of anthropogenic climate change.

Dr Kevin Trenberth, who plays a leading role in developing the IPCC science, provided all the inspiration required in a recent paper entitled 'Framing the way to relate climate extremes to climate change.'[10] His essential message was that to ask 'to what degree climate change contributed to an event' was actually asking the wrong question. His way of framing the climate debate is that every event is influenced by climate change.

The implication of this is that because the science cannot answer the question, it doesn't have to—scientists should just claim everything is a sign of climate change. None too surprisingly, the Australian Climate Commission heeded this call which was most clearly advocated in its recent 'Angry Summer' report. It concluded that everything that happened that summer was due in part to climate change. This is the kind of science many might prefer to the real thing—a science where one doesn't actually have to do anything to justify one's claims. In reality, Trenberth's framing of the climate science debate has little to do with science—it is merely advocacy for a catastrophic future outlook. Above all, it represents an intellectually weak approach to science from those that lead it.

One very unfortunate event followed just a week after the wide-spread floods of 2010-11—a paper was published in the journal *Nature* which claimed to have linked increases in rainfall to anthropogenic climate change.[11] This was heralded across the Australian Broadcasting Commission as a significant result. Expert commentators were sought to evaluate its meaning. To paraphrase, one climate scientist announced that 'it was published in *Nature*, so it must be right'; another claimed that 'we already knew this, so it only confirms what we already thought.' Such comments could have been as easily made without even bothering to read the paper. No critical analysis was ever provided. The unfortunate timing, coming so soon after the floods, meant that inevitably many scientists were emboldened in linking our floods to increased temperatures and consequently climate change.

The paper has gone on to enjoy great academic success, having been cited more than three hundred times in its short period of existence. There is however one rather major problem with the paper—the study never did compare the calculated rainfall probabilities against the corresponding temperature. If it had, it would have noted that there was no correlation at all between the two.

Figure 1 shows the five-year average one-day rainfall probabilities (black line) from 1951 onwards. There is a spike at the end, but no substantive evidence for a consistent trend. The grey line shows northern hemisphere temperature anomalies for the same periods. Note that between 1951 and 1975, temperature anomalies were declining, whilst half of the apparent increase in rainfall occurred. A more appropriate conclusion of the paper could have been that (i) rainfall intensities are highly variable in time, and (ii) temperature appears not to significantly influence the risk of intense rainfalls. The paper would not have been published in *Nature*, nor would it have received any of the attention that it got (and still gets).

The reality is that climate is hugely complicated and highly variable

Figure 1: Rainfall (black) and temperature (grey) anomalies 1952-2000

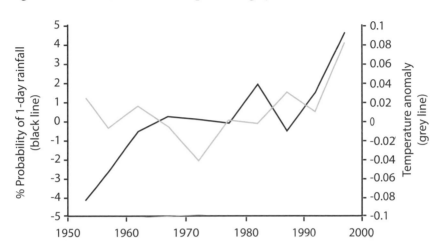

Source: S-K Min, X. Zhang, F.W. Zwiers and G.C. Hegerl, "Human contribution to more intense precipitation extremes," *Nature,* Vol. 470 (2011), 378-381

and that it is actually meaningless to attempt to identify anthropogenic climate change through climate events like floods or droughts.

So if this is the case—how does one test climate change?

The key to testing claims of catastrophic anthropogenic climate change lies in the simplicity of the mechanism itself—atmospheric carbon dioxide concentrations are rising, carbon dioxide is a greenhouse gas, and it absorbs and re-radiates longwave radiation downwards. The radiation is trapped (or at least delayed) meaning that there is more of it around and hence temperatures go up. So the simplest way to test the hypothesis is to measure the planet's energy balance, in particular downward longwave radiation.

In search of the signal of climate change, a recent NASA study has attempted to do just this.[12] The method itself is rather complex involving multiple satellite remote-sensing platforms, radiative modelling, and a whole raft of assumptions and uncertainties. That said, the results are viewed as the most meaningful way to estimate the role and context of

CO_2 in the atmosphere. Figure 2 shows the estimated global fluxes of the surface energy balance.

The 24-year record of longwave radiation is shown in the middle graph—what is immediately apparent is that there are large year to year variations. Major positive variations are linked to particularly strong El Niño events. Underlying this variability, one can perceive a relatively minor general increase in longwave flux, however it is small relative to the naturally occurring variability within the series. It is worthwhile to note at this point that atmospheric carbon dioxide rose by approximately 25 per cent over this period—the longwave response to this (the actual mechanism of climate change) appears rather muted if not entirely underwhelming.

Of even greater significance is the shortwave radiation (shown in the top graph). This is the amount of solar radiation (sunshine) received at the surface and so is primarily influenced by global cloudiness but also dust and aerosols (note the two to three-year decline in shortwave following the 1991 eruption of Mt. Pinatubo). There are a number of aspects of Figure 2 that are of importance to note.

First of all, the variability in the incoming shortwave is much greater than the variability in the longwave. In fact, it is approximately twice as large. Consequently, the shortwave variability is clearly more important to the overall energy balance. Also important to note is the decadal timescales of variability—in particular, there appears to have been a substantial declining trend since around 2000.

Finally, the shortwave variability shows no apparent correlation with the longwave variability. Consequently, changes in longwave radiation do not appear to have any influence on global cloudiness which has a larger influence on the overall energy balance on decadal timescales. In Figure 2, the global net radiation anomalies are shown in the bottom graph in comparison to the global temperature record shown in the top graph. The net imbalance appears to have been negative since 2000, during

Figure 2: Surface radiation anomalies—solar/shortwave (top), longwave (middle) and net balance (bottom)

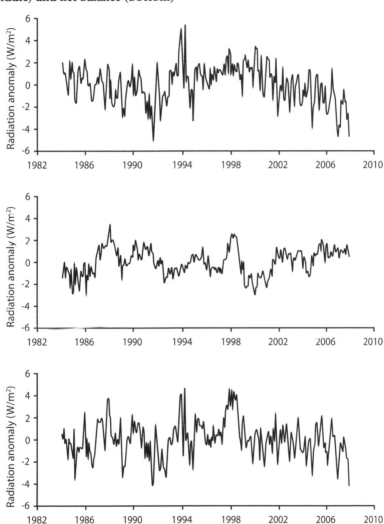

Source: P. W. Stackhouse, S. K. Gupta, S. J. Cox, T. Zhang, J. C. Mikovitz, L. M. Hinkelman, "24.5-Year Surface Radiation Budget Data Set Released," *GEWEX News* Vol. 21 No. 1 (February 2011), accessed 10 July 2014, http://www.gewex.org/images/Feb2011.pdf. Note that shortwave variability is greater than longwave radiation, demonstrating cloud variability dictates energy balance.

which time global temperatures have not risen.

This then begs the question—what does cause these variations in the Earth's energy balance? There have been a range of explanations offered to explain away this recent hiatus in temperatures, as well as the previous hiatus that occurred between 1945 and 1975.

In the first instance, many climate scientists have simply refused to accept that these periods exist in anything other than a random statistical sense—natural year-to-year variability gives statistical anomalies over short periods. Others have suggested that it is global emissions of aerosols, which have a net cooling effect, ironically especially those emitted from burning coal. In this narrative, the 1945-1975 cooling was caused by the post-war boom, whilst the most recent hiatus is due to the recent rise of Chinese coal burning.[13]

More recently, the favoured narrative is that the global temperature has been relatively static because the additional heat is being trapped in the oceans instead. A recent paper, again published in *Nature*,[14] suggested that this was in large part due to a natural multi-decadal variability, termed the Pacific Decadal Oscillation (PDO)[15] or the Inter-decadal Pacific Oscillation (IPO),[16] and associated with El Niño and La Niña behaviour. The narrative is that natural processes mean that the extra heat goes into the oceans, but that at some time soon, the heat will return with a vengeance.

There are a number of problems with these claims—importantly, the trapping of the heat is inferred from models, not from observations. The NASA data seems to suggest that it is not excess energy being trapped in the ocean, but rather that the change in the IPO/PDO has led to increased cloudiness, hence a reduction in the incoming shortwave radiation around the year 2000 (as seen in Figure 3). This would mean that there is no excess to be trapped, whether in the ocean or anywhere else. Not surprisingly, the recent study did not consider, let alone evaluate this.

Perhaps the greatest problem with this recent result is that in many

Figure 3: Monthly net surface energy balance and global temperature anomalies

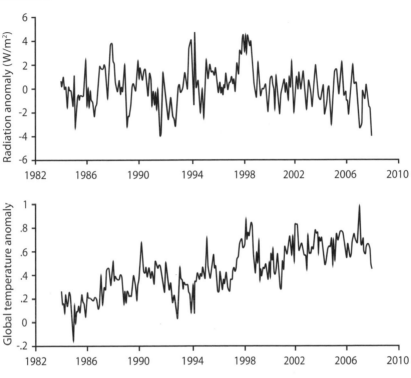

Source: HadCRUT4

ways it is not recent at all—the climate community has known about this mode of climate variability since the late 1990s. Since 2000, I myself (amongst others) have published extensively on the IPO/PDO and its strong links to multi-decadal flood and drought risk in Australia (i.e. 'the rainfall we had in the 1950s, 1960s and 1970s'). The papers also noted its link to variable periods of global warming and cooling, pointing out that the climate models could not simulate it and that perhaps we could expect a return to a cooling period at some point in the near future.[17] Little did we know that in 2000 it had already begun.

This is the kind of science I prefer. One which develops an insight and makes a prediction which can be evaluated in terms of what subsequently occurs. The majority of climate scientists, blindly relying on climate models, fail in their predictions. They adapt to this by explaining why the reality is different from the models. Theirs is not science, but the continual re-writing of a narrative and always after the event—there is always a story to explain why their models fail, but also why we should all expect catastrophic climate change nonetheless.

Karl Popper, the great philosopher of science, observed and defined this quite clearly. A theory has to be tested—he called this falsification. This requires making a prediction and then testing it. Rather than rejecting a failed theory, scientists might adapt their theory to include the anomaly in the testing. This he called fortification—if the theory fails, adapt it, fortify it until it fits. The problem with fortification is that you cannot keep doing it for ever—eventually there are too many patches and band-aids to support it.

The most important thing to note is that this lack of warming was not predicted by climate models. This is because climate models are built predicated on the assumption that increased CO_2 leads to increased water vapour and together, through a positive feedback loop, lead to runaway global warming because of a runaway increase in downward longwave radiation.

The NASA energy balance data appear to lead us to reject this hypothesis. The temperature data themselves lead us to reject climate models as accurate simulators of the global climate. Instead climate scientists speculate about flood and droughts, mostly in ignorance. They also speculate about everything else that could be impacted by climate.

The fervour with which some speculate following natural climate disasters is in stark contrast to the ability to attribute these events to atmospheric CO_2 concentrations. In their frustration, many now claim all events are influenced (we just don't know by how much).

A more scientific approach is to directly test the hypothesis by directly measuring the proposed mechanism, specifically, the energy balance of the planet. The results to date, whilst only suggestive, do point to a much more conservative role of CO_2 in influencing the energy balance than the climate models predict. Consequently, predictions of catastrophic climate change (above and beyond natural catastrophic climate variability) seem somewhat premature, if not grossly exaggerated. Above all, there is no insight gained by speculating on individual floods, droughts or even their short-term trends.

I do recall one study some years back that claimed that a species of chicken was becoming smaller as local temperatures rose. One author of the study ruefully commented to the media 'If climate change continues like this, then one day you'll be able to fit them in your pocket'. I assume that at least this scientist could see the absurdity of such speculation.

20   Extreme weather and global warming

Anthony Watts

Up until recently, the posited effects of an increased global average surface temperature were mostly limited to the direct effects of temperature itself. These included sea level rise (by thermal expansion of ocean water and ice-melt of the ice caps), shrinking glaciers, and longer, more intense heat waves, to name a few.

Most of these are somewhat distant effects for the average person. The average person living in the midwestern United States wouldn't be affected at all by sea level rise, or loss of glaciers, or the shrinking of the ice caps. That person might be affected by increased temperature and possibly an extended heat wave, but both of these are things that can be adapted to. Low-cost air-conditioners are accessible to a vast majority of the population, which wasn't the case during the dust bowl years and other heat waves of the past in America.

Essentially, these posited effects of an increased global average surface temperature just aren't much of a concern in the daily lives of many people. Even people who live in coastal zones can't detect the slow pace of sea level rise within their lifetimes, which ranges from 1.7 mm per year[1] from tide gauge measurements to about 3.3 mm per year[2] based on satellite observations. Assuming the rates hold, over a 70 year lifetime, such changes would amount to 111 mm (4.3 inches) to 231mm (9.1

inches). The rate of change is so slow as to be almost undetectable in the human experience. Likewise, the rate of global temperature change since the early twentieth century is generally agreed to be about 0.8°C (1.4°F), again so small to be almost undetectable in the human experience. In fact, if the global average surface temperature data is plotted in the same scale as a standard outdoor home thermometer, the change of the last 130 years is hardly even visible, as this graph of NASA Goddard Institute for Space Studies (GISS) data shows in Figure 1.

Since such plots generally don't get people all that concerned, for obvious reasons, most depictions of the global average surface temperature are done with temperature anomaly graphs, which magnify the change within a small range and show the change (anomaly) from a base

Figure 1: NASA GISS Global Land-Ocean Temperature Index plotted as annual average temperatures on an absolute scale similar to a liquid in glass alcohol thermometer

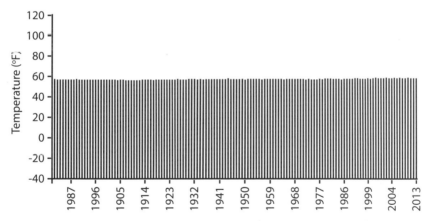

Source: Plot of NASA GISS global average surface temperature from 1880-2013, with thanks to James Sexton for conversion assistance. Data derived from "GLOBAL Land-Ocean Temperature Index in 0.01 degrees Celsius base period: 1951-1980," National Aeronautics and Space Administration, accessed July 10, 2014, http://data.giss.nasa.gov/gistemp/tabledata_v3/GLB.Ts+dSST.txt

average temperature. They also have another advantage and that is the choice of the baseline period, in the case of NASA GISS, considered the original reference source for global warming claims, they use a baseline period from 1950-1980. This just happens to be the coolest period of global temperature in the twentieth century, thus making the deviation from the 30-year average temperature. An example is shown in Figure 2.

With an anomaly plot, the scale is highly magnified compared to the scale of Figure 1, and shows the smallest of variances within the range of approximately 1°C (1.8°F) in the vertical axis of the graph. From such depictions, claims that global warming is occurring rapidly are backed up by the steep slope of the visual depiction, but in reality, as shown in Figure 1, global temperature has been remarkably stable for over a century, with

Figure 2: Global mean land-ocean temperature change (anomaly) from 1880–2013, relative to the 1951–1980 average temperature

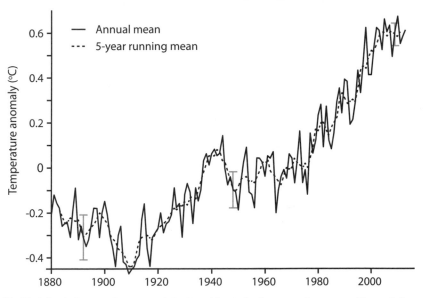

The black line is the annual average and the dotted line is the 5-year running average. The vertical bars show uncertainty estimates.

Source: NASA GISS, available at http://data.giss.nasa.gov/gistemp/graphs_v3/.

variances that are small when compared to the scale of human temperature experience. So, concern about the problem doesn't sell well in the broad court of public opinion. However, as you can see in Figure 2 at the upper right, the temperature plot has stopped rising, and for over a decade has remained close to the 0.6°C line. This period of hiatus in temperature rise has become known as 'the pause.'

This is part of the reason that terminology used to describe the phenomenon has shifted from 'global warming' to 'climate change,' as climate change can be used as a catch-all phrase without the need to address the inconvenient pause in warming.

A study by Yale University concluded that the shift of terminology from 'global warming' to 'climate change' has backfired; as it has a milquetoast connotation in the minds of the public.[3] According to the study, the phrase 'climate change' tends to be associated with unusual, but not necessarily terrifying weather events.

With 'the pause' in temperature looming large, and with a passive response by the public to the phrase 'climate change,' those who have an interest in pushing the issue tried a new tactic; they tried to connect global warming to everyday weather events.

In September 2010, the White House declared that the phrase should now be 'global climate disruption'. The phrase 'global climate disruption' was created out of thin air by John Holdren, director of the White House Office of Science and Technology Policy himself, as there were no studies or reports prior to the one he wrote in 2007 in which he declared:

> 'Global warming' is now a phrase on everyone's lips—it has more than 50 million hits on Google. Its combination with biodiversity—the variety of life on Earth—gets more than a million hits, barely 15 years since Peters and Lovejoy convened the first meeting on the subject. The phrase is appealing, but seriously misleading. Earth is experiencing a rapid global disruption to its climate, one of considerable physical complexity.[4]

This was further emphasised in a 2012 paper by Dr. James Hansen of NASA GISS (who also was the keeper of the global temperature record plotted in Figure 1 and Figure 2) who said:

> Our analysis shows that it is no longer enough to say that global warming will increase the likelihood of extreme weather and to repeat the caveat that no individual weather event can be directly linked to climate change. To the contrary, our analysis shows that, for the extreme hot weather of the recent past, there is virtually no explanation other than climate change. The deadly European heat wave of 2003, the fiery Russian heat wave of 2010 and catastrophic droughts in Texas and Oklahoma last year can each be attributed to climate change. And once the data are gathered in a few weeks' time, it's likely that the same will be true for the extremely hot summer the United States is suffering through right now.[5]

But this impression is untrue. These events and others like them almost certainly would have occurred on their own (i.e. naturally). Climate change may have added a pinch of additional heat, but it almost certainly did not create these events out of thin air.

Hansen pushes his impression with an analogy of 'climate dice.' The idea is that anthropogenic greenhouse gas emissions have 'loaded' the dice towards extreme warmth, so now when Mother Nature rolls the dice for summer weather, there is better chance of rolling a heat wave, or an overall hot summer—events discreet from events that were contained on the unloaded dice.

But Hansen's hot summers are not new discrete events at all. Instead, they are the naturally occurring hot summers with a few extra degrees added to them. The extra couple of degrees push some summers over an arbitrarily defined threshold temperature above which Hansen then classifies them as being 'extreme.'

Hansen's threshold between a 'normal' summer and an 'extreme' summer has no physical meaning—instead it is rooted in statistics. While certainly some temperature thresholds exist that have physical meaning—like the 32°F, the freezing/melting point of water/ice—none exist in the range of temperatures which characterise summer across much of the globe. Whether or not the average summertime temperature is greater or less than some arbitrary value is of little practical significance.[6]

So how do climate advocates get people to care about such things that they can hardly see or feel? They make it *personal*, by trying to connect virtually every weather event to global warming/climate change/global climate disruption, because weather is something that is universal to the human experience and detectable on a daily basis. The idea that weather can affect you is borne out in the daily news throughout history; floods, thunderstorms, tornadoes, hurricanes, blizzards, droughts and heat waves are a shared part of the human experience globally via pain, suffering, and property destruction.

What better way to make people concerned about small, virtually imperceptible, century-scale changes in temperature than by saying that these changes directly affect the daily weather, making it more extreme?

This fear-mongering speaks to the human psyche, for mankind has long feared the effects of severe weather and the primal urge is to seek shelter from it and to avoid it whenever possible. It is a basic survival instinct learned from experience, but contrary to the lack of any learned experience from a century of virtually undetectable global temperature change.

Why it seems that extreme weather is 'getting worse' when the data shows otherwise

With the help of the electronic media, many, if not all, of the extreme weather events we see globally on a day to day basis are etched into the minds of people simply by the act of watching news broadcasts, reading newspapers, listening to radio, getting Twitter or Facebook alerts, receiving SMS messages on cellphones, or reading web pages. Live television news broadcasts gravitate towards what is action-packed and exciting, which tends to cater to the viewer's emotions, rather than address the factual content. This is essentially the *modus operandi* of the electronic media, and particularly, television news. The goal is to capture eyes and ears, and to keep them engaged.

In 2011, Bouziotas et al. presented a paper on flood trends that concluded:

> Analysis of trends and of aggregated time series on climatic (30-year) scale does not indicate consistent trends worldwide. Despite common perception, in general, the detected trends are more negative (less intense floods in most recent years) than positive. Similarly, Svensson et al. (2005) and Di Baldassarre et al. (2010) did not find systematical change neither in flood increasing or decreasing numbers nor change in flood magnitudes in their analysis.[7]

Note the emphasised phrase 'Despite common perception'. That 'common perception' is central to the theme of 'extreme weather is getting worse' which was started by John P. Holdren in his 2007 paper on 'global climate disruption'.

Like Holdren, many people who ascribe to doomsday scenarios related to anthropogenic global warming seem to think that extreme weather is happening more frequently. Most people are not educated in the history of television technology, web technology, and mass media, and without that background, it is easy to miss the central lesson about why the false perception exists of an increase in extreme weather today.

Weather appears more extreme today, not because it is, but because we hear about it nearly instantly, and such reports saturate the electronic media within minutes of occurrence.

Compare the reach and speed of communications and news reporting at the beginning of this timeline to the reach and speed of communications and news reporting technology around the beginning of the twentieth century. Then compare that to the beginning of the twenty-first century. Compare again to what we've seen in the last ten years.

With such global coverage, instant messaging, and internet enabled phones with cameras now, is it any wonder that nothing related to severe weather or disaster escapes our notice any more? Certainly, without considering the technological change in our society, it would seem as if severe weather events and disasters are becoming much more frequent.

To borrow and modify a famous phrase from James Carville: it's the technology, stupid.

All of these advances in communication speak to the phrase, 'despite common perception,' which was highlighted at the beginning of this section. The speed of weather tracking and the communications technology curve aids in our 'common perception' of extreme weather events, but the reality of extreme weather frequency though, is actually quite different. While we may see more extreme weather on a daily, monthly, and yearly basis, that happens only because there are millions more eyes, ears, cameras and networks than ever before.

Extreme weather was always there, but up until recently in human history there was no way to record it and share it quickly. Now almost anyone with a camera enabled cell phone can report on extreme weather from nearly anywhere on the globe and have it in the hands of television networks and internet news sites within minutes of occurrence.

Conclusion

In the IPCC's Fifth Assessment report, *Summary for Policymakers* (SPM), there is no mention at all of tornadoes or hurricanes in the extreme weather events section. They give low confidence to tropical storm activity being connected to climate change, and don't mention mesoscale events like tornadoes and thunderstorms at all. Similarly, they give low confidence to drought and flood attribution.

They've only talked about heat waves and precipitation events being connected. From the SPM:

Changes in many extreme weather and climate events have been observed since about 1950 … It is very likely that the number of cold days and nights has decreased and the number of warm days and nights has increased on the global scale. It is likely that the frequency of heat waves has increased in large parts of Europe, Asia and Australia. There are likely more land regions where the number of heavy precipitation events has increased than where it has decreased. The frequency or intensity of heavy precipitation events has likely increased in North America and Europe. In other continents, confidence in changes in heavy precipitation events is at most *medium*.[8]

This is consistent with what was reported in the *IPCC Special Report on Extremes* (SREX).

There is medium evidence and high agreement that long-term trends in normalized losses have not been attributed to natural or anthropogenic climate change … The statement about the absence of trends in impacts attributable to natural or anthropogenic climate change holds for tropical and extratropical storms and tornados … The absence of an attributable climate change signal in losses also holds for flood losses.[9]

This lack of attribution of severe storms to 'man-made climate change' in AR5 contradicts the claims of Hurricane Sandy, tornado outbreaks, floods, and other media sensationalisms about imagined connections with climate change.

In addition to two IPCC reports making no connections between extreme weather and climate, we have *Nature*'s editorial in 2012 saying:

Better models are needed before exceptional events can be reliably linked to global warming …

To make this emerging science of 'climate attribution' fit to inform legal and societal decisions will require enormous research effort.[10]

When the journal *Nature* says that there is no reliable linkage and 'enormous research effort' is needed to link extreme weather as a byproduct of global warming, climate change, or climate disruption, you know that no matter what you want to call it, it is a dead issue with true science at the moment, and the value of such wild claims trying to link extreme weather with climate exists only as a recruitment tool for climate activists and zealots.

21 False prophets unveiled
Andrew Bolt

Not once when boarding a ship have I thought of ducking into the bridge to give the captain pre-voyage tips on navigation. He's the expert.

Still, if I'd booked for Noumea only to dock in Karachi I would know this: whatever the captain's credentials, he'd goofed. We weren't where he'd promised to take us, and him screaming that he understands the sailing business better than do I, while true, won't cut it. Next time I'm flying.

I thought such a principle was so obvious that all laymen would consult it in every contact with any professional. Your new implant falls out? Sack the dentist. Your extension falls down? Sue the builder. Clowns, both of them.

But in one tiny yet catastrophically expensive field of human endeavour this law seems to have been suspended for a decade or two. Yes: climate science. This is the science where one plus one can equal three one day and six the next—yet never may the layman question the expert at the blackboard, or the shill demanding a few billion to make the sum equal no more than two.

This must change, and I believe finally is. The tyranny of the experts is now crumbling. The common sense of the layman is at last being restored.

Hey, didn't you guys say it would never flood? Then what's this stuff that's washed my car down my Queensland street? Hmm, didn't you also predict runaway warming? So why these sixteen years of non-warming? Think I'll get me some new experts.

See, after more than a decade of scares we are now getting the years of the busted predictions—that 'Er, this isn't Noumea' moment at the top of the gangplank. We are waking up to nearly two decades of wild predictions by experts who now look like geese to the sane, despite still being hailed as gurus by the last holdouts of the global warming faith—in Parliament, in universities and in the great fortress of our State broadcaster.

But how hard the politicians have worked to make us trust the very people we should have questioned before we handed over $8 billion a year in carbon taxes to splurge on making no difference at all to the weather.

Think back to 2007—the year Professor Tim Flannery, the professional alarmist, had warned Sydney, Brisbane and Adelaide could run out of water.

In Sydney, Prime Minister Kevin Rudd, with a huge dam nearly two thirds full on the other side of the city, instructed Australians to ignore the few of us now pointing out emperor Flannery had no dry clothes. Trust the experts, he demanded, and not those amateurs pointing out they'd docked in Karachi.

'The truth is that the do-nothing climate change sceptics offer no alternative official body of evidence from any credible government in the world,' Rudd said in his Lowy Institute speech. Then came my exciting name-check in Rudd's list of those-who-must-not-be-heeded: 'Malcolm [Turnbull], Barnaby [Joyce], Andrew, Janet [Albrechtsen]—stop gambling with our future. You've got to know when to fold 'em—and for the sceptics, that time has come.' Oh, really?

Fast forward to 2011, and it was the turn of Rudd's successor, Julia Gillard, to demand Australians listen only to experts such as the Bureau of Meteorology, which three years earlier had actually predicted the long drought was our new normal and 'perhaps we should call it our new climate'.

We'd since had two years of devastating floods in Queensland, as well as floods in New South Wales and Victoria, yet Gillard still insisted Australians trust the Bureau's alarmists. 'I ask, who would I rather have on my side?' she said, 'Alan Jones, Piers Akerman and Andrew Bolt? Or the CSIRO, the Australian Academy of Science, the Bureau of Meteorology, NASA, the US National Atmospheric Administration, and every reputable climate scientist in the world?'

Every reputable climate scientist in the world, that is, except those such as Professor Richard Lindzen and Dr Roy Spencer who warned and warned again that the world simply wasn't warming as the warmists predicted.

But arguments are settled by evidence, not a show of hands. As Albert Einstein reportedly said in response to the book *One hundred authors against Einstein*, disputing his new Theory of Relativity: Why 100 authors? 'If I were wrong, then one would have been enough.' One fact can disprove even 1000 experts.

And that is what the layman must never forget: truth is decided by evidence, not qualifications or a show of hands. True, the average climate scientist is far better equipped than the average layman to understand where the truth lies, but facts sometimes speak so loudly that even the greatest scientist can be doubted.

These include the Karachi moments—when what 'the experts' predicted has been contradicted by time and facts. When we learned that whatever map a climate scientist was reading, it's brought us to the wrong place.

I've listed below twenty of those moments that give us not just a reason to doubt the alarmists, but a duty. But let's first look at the biggest prediction of them all: that the planet will heat dangerously and fast. This is the meta-prediction upon which all the other predictions hang—about fast-rising seas, failing crops, melting ice caps, permanent droughts, worse epidemics and mass extinctions.

In 2001, the United Nations' Intergovernmental Panel on Climate Change (IPCC) predicted a huge and rapid rise in global surface air temperatures thanks to man's emissions, 'Projections using these scenarios in a range of climate models result in an increase in globally averaged temperatures of 1.4 to 5.8 degrees Celsius over the period 1990-2100.'

A rise in temperatures of up to six degrees this century? Cue terrifying scenarios.

Professor James Lovelock, the world-famous Gaia guru, warned in 2006 we'd be wiped out: 'Before this century is over billions of us will die and the few breeding pairs of people that survive will be in the Arctic where the climate remains tolerable.'

Professional alarmist Professor Tim Flannery in 2007, at the height of the warming scare and four years before being appointed Australia's Chief Climate Commissioner, said the certainty of big rises in temperature meant we faced mass extinctions. 'Three degrees will be a disaster for all life on earth ... It could be worse than this—there's a 10 per cent chance of truly catastrophic rises in temperatures, so we're looking at six degrees or so.'

Journalist Mark Lynas was inspired to write the bestseller *Six Degrees*, predicting an apocalypse. This in turn inspired Channel Nine's *A Current Affair* in 2007 to show huge balls of fire smashing into cities by the end of the century, as a narrator intoned, 'This is six degrees. Flash floods. Gas and methane fireballs racing across the globe with the power of atomic bombs. Life on earth ends in apocalyptic storms.'

But a curious thing happened. As the years went by, the world's atmospheric temperatures stubbornly refused to rise. In fact, by 2014 that pause in the warming had lasted sixteen or seventeen years, depending on which data sets are consulted.

The leaked *Climategate* emails showed growing alarm at this pause among the very IPCC climate scientists most responsible for the warming scare—an alarm they kept amongst themselves. For example, Dr. Phil Jones, director of the Climate Research Unit at the University of East Anglia, privately confessed in 2005, 'The scientific community would come down on me in no uncertain terms if I said the world had cooled from 1998. Okay it has but it is only seven years of data and it isn't statistically significant ... '

Dr. Kevin Trenberth in 2009 wondered 'where the heck is global warming?' adding, 'The fact is that we can't account for the lack of warming at the moment and it is a travesty that we can't.' Jones in 2009 counselled against panic, 'Bottom line: the"no upward trend" has to continue for a total of 15 years before we get worried.' The trend of no warming has since exceeded that fifteen-year mark.

But whatever their private doubts, the warmists kept denying publicly that their predictions really were bust. In 2008, for instance, Britain's warmist Met Office insisted 'we expect that warming will resume in the next few years'. It hasn't, of course.

In 2012, Professor Matthew England of the NSW Climate Change Research Centre even appeared on the ABC's *Q&A* to rebuke former Industry Minister Nick Minchin for simply saying what the Met had already admitted—that the planet's atmosphere had not warmed since 1998 by any statistically significant amount.

'What Nick just said is actually not true,' England insisted. 'The IPCC projections from 1990 (of steady rises) have borne out very accurately.' England later even accused sceptics of 'lying that the IPCC projections are overstatements'.

So imagine my surprise when, just two years later in 2014, England admitted there had been a 'hiatus' and 'plateau in global average temperatures' after all. Indeed, Dr Roy Spencer, who runs the University of Alabama at Huntsville global temperature data collated by NASA's Aqua satellite, says the climate models used to predict man-made warming of up to six per cent this century have so far 'failed miserably'. More than 95 per cent of the 90 climate models had 'over-forecast the warming trend since 1979'. He added: 'Whether humans are the cause of 100% of the observed warming or not, the conclusion is that global warming isn't as bad as was predicted.'

Good heavens. So many flawed predictions, then, from so many experts who for years shouted down sceptics with the cry, 'Respect the science! Don't argue: 97 per cent of scientists agree!' And meanwhile, behind the scenes, the frantic search is on to find their missing heat. It's hiding in the oceans! No, it's suppressed by aerosols! No, wait—it's masked by natural factors that will soon give way and then … watch out!

Except the bottom line is that many experts no longer quite agree with what they once predicted about our rising atmospheric temperatures, and our apocalyptic fever is slowly waning. In 2012 even Lovelock, the Gaian catastrophist, admitted he'd been 'alarmist' and so had other leaders of the warming alarm, including Flannery and Al Gore.

'The problem is we don't know what the climate is doing. We thought we knew 20 years ago. That led to some alarmist books—mine included—because it looked clear-cut, but it hasn't happened,' Lovelock said. 'The world has not warmed up very much since the millennium. Twelve years is a reasonable time … It (the temperature) has stayed almost constant, whereas it should have been rising.'

Had Lovelock been our captain, we'd be sailing to the Arctic. Except now he says we'll be landing just where we started from, after all. Someone goofed, and we should duck into the bridge to say so.

Prediction: our drought is permanent and our cities may run out of water

In 2005, Flannery, then Climate Council head, predicted: 'We've seen just drought, drought, drought … If you look at [Sydney's] Warragamba catchment figures, since '98 the water has been in virtual freefall, and they've got about two years of supply left ... They (the changes) do seem to be of a permanent nature.'

In fact, the drought then broke. Warragamba, Sydney's main dam, has since filled to overflowing.

In 2009, Flannery predicted: 'The soil is warmer because of global warming and the plants are under more stress and therefore using more moisture. So even the rain that falls isn't actually going to fill our dams and our river systems ...'

In fact, Queensland, NSW and Victoria have since suffered severe floods. Dams in Brisbane, Sydney and Canberra have all filled.

In 2007, Flannery predicted: 'In Adelaide, Sydney and Brisbane, water supplies are so low they need desalinated water urgently, possibly in as little as 18 months.'

In fact, both Sydney and Brisbane's dams were more than 85 per cent full as of March 2014. Adelaide's catchments were 62 per cent full. Sydney and Brisbane have mothballed their desalination plants.

In 2009, Bertrand Timbal, a Bureau of Meteorology climatologist, predicted: 'The rainfall we had in the 1950s, 60s and 70s was a benchmark, but we are just not going to have that sort of good rain again as long as the system is warming up.'

In fact, the Bureau has since declared 2010 and 2011 Australia's wettest two-year period on record'.

Prediction: the Great Barrier Reef is being wiped out by warming

In 1999, Professor Ove Hoegh-Guldberg, a Queensland University reef expert and an Intergovernmental Panel on Climate Change lead author, predicted warming would so heat the oceans that mass bleaching of the Reef would occur every second year from 2010.

In fact, the Reef's last mass bleaching occurred in 2006.

In 2000 Hoegh-Guldberg claimed 'we now have more evidence that corals cannot fully recover from bleaching episodes such as the major event in 1998' and 'the overall damage is irreparable'.

In fact, Hoegh-Guldberg admitted in 2009 he was 'overjoyed' to see how much the reef had recovered, and the Australian Institute of Marine Science says 'most reefs recovered fully'.

In 2006, Hoegh-Guldberg warned high temperatures meant 'between 30 and 40 per cent of coral on Queensland's Great Barrier Reef could die within a month'.

In fact, Hoegh-Guldberg later admitted this bleaching had 'a minimal impact'.

In 2011, Hoegh-Guldberg predicted a 'large-scale mortality' of reef-building corals on West Australian reefs from Shark Bay to Exmouth within three months.

In fact, Hoegh-Guldberg later admitted the famous Ningaloo Reef, the largest there, actually 'had a narrow escape'.

Prediction: global warming is causing massive rises in sea levels, drowning islands

In 2007, Professor Mike Archer, dean of science at the University of NSW, said: 'Forget Venice; I mean we're talking about sharks in the middle of Sydney' because the seas would rise '100 metres'. The ABC's chief science presenter, Robyn Williams, agreed 'it is possible, yes' that this would occur before the end of this century.

In fact, sea level rises for the past twenty years have averaged just 3.2 millimetres a year, according to the University of Colorado monitoring—or 30 centimetres a century. Sea levels have slowly risen since 1880, well before human influence on the climate is said to have become significant.

In 2006, warmist alarmist Gore claimed in his film *An Inconvenient Truth* that seas were rising so fast 'that's why the citizens of these Pacific nations have all had to evacuate to New Zealand'.

In fact, in a British court case, Justice Michael Burton found 'there is no evidence of any such evacuation having yet happened.'

In 2013, Labor Foreign Minister Bob Carr claimed the Pacific island of Kiribati 'is at the front line of climate change' and 'unless action is taken, Kiribati will be uninhabitable by 2030 as a result of coastal erosion, sea level rise and saltwater intrusion into drinking water'.

In fact, the 1993-2011 sea level trend data from Tarawa atoll, part of Kiribati, shows no rise in sea level. The most populous atoll of Kiribati—the tiny islet of Betio—has increased in size by a third over the past 60 years. In 2010 an Auckland University survey noted that 86 per cent of 27 Pacific islands studied—including Kiribati and Tuvalu—had grown or stayed the same size over the past twenty to 60 years.

Prediction: our drought will be permanent

In 2003, Melbourne warmist scientist David Karoly claimed, 'drought severity in the Murray Darling is increasing with global warming'.

In fact, the rains returned, the Murray-Darling flooded and the Climate Commission in 2011 admitted, 'it is difficult from observations alone to unequivocally identify anything that is distinctly unusual about the post-1950 pattern [of rainfall]'.

Prediction: sea ice is vanishing and the Arctic will be ice-free

In 2008, Flannery asked people to imagine 'a world five years from now, when there is no more ice over the Arctic', and Gore predicted 'the entire north polar ice cap will be gone in five years'. Ted Scambos, of the US Snow and Ice Data Centre, told the ABC there was 'a very strong case that in 2012 or 2013 we'll have an ice-free (summer) Arctic'.

In fact, at the height of the summer melt last year, the Arctic was still covered by six million square kilometres of ice, more than in the previous three years.

In 2009, Gore predicted the Antarctic would melt away, too: 'They're seeing the complete disappearance of the polar ice caps ... ' Professor Chris Turney of the NSW Climate Change Research Centre in 2013 led an expedition to Antarctica, claiming 'Sea ice is disappearing due to climate change … '

In fact, the seas around Antarctica had more ice cover in 2013 than seen since satellite records started in the late 1970s. NASA says sea ice cover in Antarctica has grown 1.5 per cent a decade for several decades. Turney's expedition got trapped in sea ice and had to be rescued.

Prediction: there will be no snow

In 2000, Dr. David Viner of the Climatic Research Unit of Britain's University of East Anglia claimed that within a few years winter snowfall would become 'a very rare and exciting event' and 'children just aren't going to know what snow is'. In 2007, Sir John Houghton, former head of Britain's Met Office, said 'less snow is absolutely in line with what we expect from global warming.'

In fact, five of the northern hemisphere's six snowiest winters in the past 46 years have occurred since Viner's prediction, according to Rutgers University Global Snow Lab numbers. Over two-thirds of the contiguous USA were covered with snow in the winter of 2013/14.

Prediction: we are not going through a plateau or pause in global temperatures

In 2007, Britain's Met Office said, 'By 2014 we're predicting it will be 0.3 degrees warmer than 2004'.

In fact, the Met Office data for 2013 confirmed there had been no statistically significant rise in global atmospheric temperatures for at least sixteen years.

In 2012, Professor Matthew England, a University of NSW climate scientist, claimed there was no hiatus in global warming and sceptics claiming that the warming was lower than predicted by the IPCC were 'lying'.

In fact, in 2014 England admitted there was a 'plateau in global average temperatures', after all. Climate scientist Professor Judith Curry told the US Congress this year: 'For the past 16 years, there has been no significant increase in surface temperature ... The IPCC does not have a convincing or confident explanation for this hiatus in warming.'

Prediction: global warming will cause starvation

In 2001, the IPCC predicted global warming would 'affect wheat and, more severely, rice productivity in India'. In 2012 the Australian Conservation Foundation claimed, 'it will be less and less likely that we can feed the human population if climate change continues on its present trajectory'.

In fact, since 1960, global wheat and rice production has tripled, and corn production is almost five times higher. Record harvests were recorded over the past decade, including in India.

Prediction: hurricanes and cyclones will get worse

In 2006, Gore, in *An Inconvenient Truth*, claimed Hurricane Katrina was evidence of global warming, adding, 'We have seen in the last couple of years, a lot of big hurricanes.' In 2011, Greens leader Bob Brown claimed Cyclone Yasi was caused by the coal mining industry because 'it's the single biggest cause—burning coal—for climate change'.

In fact, in 2013 the IPCC admitted it had 'low' confidence in claims of 'large scale changes in the intensity of extreme extratropical cyclones since 1900' and said the number of cyclones and added that since hurricanes reaching land had fallen 'over periods of a century or more, evidence suggests slight decreases in the frequency of tropical cyclones making landfall in the North Atlantic and the South Pacific.' The Bureau of Meteorology reports fewer cyclones reaching Australia.

Prediction: global warming will cause bigger hailstones

In 2007, Professor Ross Garnaut, the Rudd Government's chief global warming adviser, ordered a steel roof for his Melbourne home, telling his local council global warming would produce 'severe and more frequent hailstorms'.

In fact, the IPCC last year admitted, 'There is low confidence in observed trends in small-scale severe weather phenomena such as hail and thunderstorms.'

References

Chapter 2
Why climate models are failing

1. K. Popper, *Conjectures and Refutations* (New York: Routledge & Kegan Paul, 1963), 43-86.

2. J. A. Francis and S. J. Vavrus, "2012: Evidence linking Arctic amplification to extreme weather in mid-latitudes," *Geophysical Research Letters*, Vol. 39, No. 6 (2012).

3. E. Barnes, "Revisiting the evidence linking Arctic Amplification to extreme weather in the midlatitude," *Geophysical Research Letters*, Vol. 40, No. 17 (2013), 4734-4739; T. Ballinger, M. J. Allen and R. V. Rohli, "Spatiotemporal analysis of the January Northern Hemisphere circumpolar vortex over the contiguous United States," *Geophysical Research Letters*, Vol. 41, No. 10 (2014), 3602-3608.

4. C. P. Morice, J. J. Kennedy, N. A. Rayner and P. D. Jones, "Quantifying uncertainties in global and regional temperature change using an ensemble of observational estimates: The HadCRUT4 data set," *Journal of Geophysical Research: Atmospheres*, Vol. 117, No. D8 (2012).

5. See, for example, P. C. Knappenberger and P. J. Michaels, "Policy Implications of Climate Models on the Verge of Failure" (Paper deliv-

ered at the American Geophysical Union Science Policy Conference, Washington, D. C.June 24-26, 2013).

6.   Morice et al, "Quantifying uncertainties," as above.

7.   See P. C. Knappenberger, P. J. Michaels, J. R. Christy, C. S. Herman, L. M. Liljegren, J. D. Annan, "Assessing the consistency between short-term global temperature trends in observations and climate model projections" (Paper delivered at the Third Santa Fe Conference on Global and Regional Climate Change, Santa Fe, New Mexico, 30 October-4 November 2011); Knappenberger and Michaels, "Policy Implications of Climate Models on the Verge of Failure," as above.

8.   G. Paltridge, "Climate change's inherent uncertainty," *Quadrant*, Vol. 58 No. 1-2 (2014).

9.   T. Kuhn, *The structure of scientific revolutions* (Chicago: University of Chicago Press, 1962).

10.   Ibid, 64.

11.   R. S. Lindzen and Y. S. Choi, "On the observational determination of climate sensitivity and its implications," *Asia-Pacific Journal of Atmospheric Science*, Vol. 47, No. 4 (2011), 377-390.

12.   Ibid, 377.

13.   A. Schmittner, N. M. Urban, J. D. Shakun, N. M. Mahowald, P. U. Clark, P. J. Bartlein, A. C. Mix and A. Rosell-Melé, "Climate Sensitivity Estimated from Temperature Reconstructions of the Last Glacial Maximum," *Science*, Vol. 334 No. 6061 (December 2011), 1385-1388.

14.   Ibid, 1385.

15.   J. D. Annan and J. C. Hargreaves, "On the generation and interpretation of probabilistic estimates of climate sensitivity," *Climatic Change*, Vol. 104, No. 3-4 (2011), 423-436.

16.   Ibid.

17.   J. H. van Hateren, "A fractal climate response function can simulate flobal average temperature trends of the modern era and the past millennium," Climate Dynamics, Vol. 40, No. 11-12 (2012), 2651-2670.

18. M. J. Ring, D. Lindner, E. F. Cross and M. E. Schlesinger, "Causes of Global Warming Observed since the 19th Century," *Atmospheric and Climate Sciences*, Vol. 2 No. 4 (2012), 401-415.

19. See previous paragraph.

20. J. C. Hargreaves, J. D. Annan, M. Toshimori and A. Abe-Ouchi, "Can the Last Glacial Maximum constrain climate sensitivity?" *Geophysical Research Letters*, Vol. 39, No. 24 (2012).

21. M. Aldrin, M. Holden, P. Guttorp, R. B. Skeie, G. Myhre and T. K. Berntsen, "Beyesian estimation of climate sensitivity based on a simple climate model fitted to observations of hemispheric temperatures and global ocean heat content," *Environmetrics*, Vol. 23 No. 3 (2012), 253-271.

22. R. W. Spencer and W. D. Braswell, "The role of ENSO in global ocean temperature changes during 1955-2011 simulated with a 1D climate model," *Asia-Pacific Journal of Atmospheric Sciences*, Vol. 50 No. 2 (2013), 229-237.

23. A. Otto, F. E. L. Otto, O. Boucher, J. Church, G. Hegerl, P. M. Forster, N. P. Gillett, J. Gregory, G. C. Johnson, R. Knutti, N. Lewis, U. Lohmann, J. Marotzke, G. Myhre, D. Shindell, B. Stevens and M. R. Allen, "Energy budget constraints on climate response," *Nature Geoscience*, Vol. 6 (2013), 415-416.

24. N. Lewis, "An objective Bayesian, improved approach for applying optimal fingerprint techniques to estimate climate sensitivity," *Journal of Climate*, Vol. 26, No. 19 (2013), 7414-7429.

25. Adrin et al., "Bayesian estimation of climate sensitivity," as above.

26. T. Masters, "Observational estimates of climate sensitivity from changes in the rate of ocean heat uptake and comparison to CMIP5 models," *Climate Dynamics*, Vol. 42 No. 7-8 (2013), 2173-2181.

27. C. Loehle, "A minimal model for estimating climate sensitivity," *Ecological Modelling*, Vol. 276 (2014), 80-84.

28. A. C. Revkin, "A Closer Look at Moderating Views of Climate Sensitivity," *The New York Times*, 4 February 4 2013, accessed 10

July 2014, http://dotearth.blogs.nytimes.com/2013/02/04/a-closer-look-at-moderating-views-of-climate-sensitivity/?_php=true&_type=blogs&smid=tw-share&_r=0.

29.   Ring et al., "Causes of Global Warming Observed since the 19th Century," as above.

30.   Spencer and Braswell, "The role of ENSO in global ocean temperature changes," as above.

31.   P. J. Michaels, P. C. Knappenberger, O. W. Frauenfelt and R. E. Davis, "Revised 21st century temperature projections, *Climate Research*, Vol. 23 (2003), 1-9.

32.   Kuhn, *The structure of scientific revolutions*, as above.

Chapter 3
Global warming, models and language

1.   N. Lewis, "An Objective Bayesian Improved Approach for Applying Optimal Fingerprint Techniques to Estimate Climate Sensitivity," *Journal of Climate* Vol. 26 No. 19 (2013), 7414-7429.

2.   R.S. Lindzen and C. Giannitsis, "On the climatic implications of volcanic cooling," *Journal of Geophysical Research* Vol. 103 (1998), 5929-5941.

3.   M. Sato, "Forcings in GISS Climate Model," NASA, last modified 13 December 2012, accessed 15 July 2014, http://data.giss.nasa.gov/modelforce/strataer/.

4.   J. Zhou and K.K. Tung, "Deducing Multidecadal Anthropogenic Global Warming Trends Using Multiple Regression Analysis," *Journal of Atmospheric Science*, Vol. 70, No. 1 (2013), 3-8.

5.   G. Roe, "Feedbacks, Timescales, and Seeing Red," *Annual Review of Earth and Planetary Sciences*, Vol. 37, No. 1 (2009), 93-115.

6.   N. J. Shaviv, "Using the oceans as a calorimeter to quantify the solar radiative forcing," *Journal of Geophysical Research*, Vol. 113 No. A11101 (2008).

7.   F. J. Wentz, L. Ricciardulli, K. Hilburn and C. Mears, "How Much More Rain Will Global Warming Bring?" *Science,* Vol. 317 No. 5835 (2007), 233.

8.   Roe, "Feedbacks, Timescales, and Seeing Red," as above, 93-115.

9.   P. M. Forster and J. M. Gregory, "The climate sensitivity and its components diagnosed from Earth Radiation Budget data," *Journal of Climate,* Vol. 19 (2006), 39-52; A. E. Dessler, "A determination of the cloud feedback from climate variations over the past decade," *Science,* Vol. 330 (2010), 1523-1527.

10.   R.S. Lindzen and Y.S. Choi, "On the determination of climate feedbacks from ERBE data," *Geophysical Research Letters* Vol. 36 (2009), L16705; R.S. Lindzen and Y.S. Choi, "On the observational determination of climate sensitivity and its implications," *Asian Pacific Journal of Atmospheric Science,* Vol. 47 (2011), 377-390; R. W. Spencer and W. D. Braswell, "On the diagnosis of radiative feedback in the presence of unknown radiative forcing," *Journal of Geophysical Research,* Vol. 115 (2010), D16109; K. E. Trenberth and J. T. Fasullo, "Global warming due to increasing absorbed solar radiation," *Geophysical Research Letters,* Vol. 36 (2009), L07706.

11.   Y.-S. Choi, H. Cho, C.H. Ho, R. S. Lindzen, S. K. Park and X. Yu, "Influence of non-feedback variations of radiation on the determination of climate feedback," *Theoretical and Applied Climatology,* Vol. 115 No. 1/2(2014), 355.

12.   Lindzen and Choi, "Determination of climate feedbacks," as above; Lindzen and Choi, "Observational determination of climate sensitivity," as above; Spencer and Braswell, "Diagnosis of radiative feedback," as above; Trenberth and Fasullo, "Warming due to increasing absorbed solar radiation," as above.

13.   Lindzen and Choi, "Observational determination of climate sensitivity," as above.

14.   R.S. Lindzen, M.D. Chou and A.Y. Hou, "Does the Earth have an adaptive infrared iris?" *Bulletin of the American Meteorological Society,* Vol. 82 (2001), 417-432.

15. Lindzen and Choi, "Determination of climate feedbacks," as above; Lindzen and Choi, "Observational determination of climate sensitivity," as above.

16. S. Edvardsson, K. G. Karlsson and M. Engholm, "Accurate spin axes and solar system dynamics: Climatic variations for the Earth and Mars," *Astronomy and Astrophysics*, Vol. 384 No. 2 (2002), 689-701; G. Roe, "In defense of Milankovitch," *Geophysical Research Letters*, Vol. 33 (2006), L24703.

Chapter 4
Sun Shunned

1. W. Soon and S. Yaskell, *The Maunder Minimum and the Variable Sun-Earth Connection* (Singapore: World Scientific Publishing, 2003), xii.

2. I thank Robert Carter, Christopher Essex, Kesten Green, Dallas Kennedy, David Legates, Christopher Monckton of Brenchley, Alan Moran, Geoff Smith and Hong Yan for help in editing and improving this article.

3. IPCC, *Principles Governing IPCC Work* (Batumi: Intergovernmental Panel on Climate Change, 2013), 1, accessed 15 July 2014, https://www.ipcc.ch/pdf/ipcc-principles/ipcc-principles.pdf.

4. Ibid., 2.

5. W. Soon and S. Lüning, "Chapter 3: Solar forcing of climate," in *Climate Change Reconsidered II: Physical Science* (Chicago: Heartland Institute, 2013), 247-348.

6. F.M.Bréon, W.Collins, J.Fuglestvedt, J.Huang, D.Koch, J.F. Lamarque, D.Lee, B.Mendoza, T.Nakajima, A.Robock, G.Stephens, T.Takemura, H.Zhang, "Anthropogenic and Natural Radiative Forcing" in *Climate Change 2013—The Physical Sciences Basis. Working Group I Contribution to the Fifth Assessment Report of the Intergovernmental Panel on Climate Change*, ed. G.Myhre, D.Shindell (Cambridge: Canbridge University Press, 2013) p659-740

7. PMOD: Physikalisch-Meteorologisches Observatorium Davos;

ACRIM: Active Cavity Radiometer Irradiance Monitor; RMIB: Royal
Meteorological Institute of Belgium. For a summary of available measurements of total solar irradiance, see N. Scafetta and R. C. Willson,
"ACRIM total solar irradiance satellite composite validation versus TSI
proxy models," *Astrophysics and Space Science*, Vol. 350 (2014), 421-442.

8.   W. Soon and D. R. Legates, "Solar irradiance modulation of equator-
to-pole (Arctic) temperature gradients: Empirical evidence for climate
variation on multidecadal timescales," *Journal of Atmospheric and Solar-
Terrestrial Physics*, Vol. 93 (2013), 45-56.

9.   F. Steinhilber, J. Beer and C. Fröhlich, "Total solar irradiance during
the Holocene," *Geophysical Research Letters*, Vol. 36 (2009), L05701;
F. Steinhilber, J. A. Abreu, J. Beer, I. Brunner, M. Christl, H. Fischer,
U. Heikkila, P. W. Kubik, M. Mann, K. G. McCracken, H. Miller, H.
Miyahara, H. Oerter and F. Wilhelms, "9,400 years of cosmic radiation
and solar activity from ice cores and tree rings," *Proceedings of the (U.S.)
National Academy of Sciences*, Vol. 109, (2012), 5967-5971.

10.   J. M. Fontenla, J. Harder, W. Livingston, M. Snow, T. Woods, "High-
resolution solar spectral irradiance from extreme ultraviolet to far infrared," *Journal of Geophysical Research*, Vol. 116 (2011), 201-08.

11.   T. W. Cronin, "On the choice of average solar zenith angle," *Journal of
the Atmospheric Sciences* (2014), accessed 14 July 2014, http://web.mit.
edu/~twcronin/www/document/Cronin2014_ZenithAngles_accepted.
pdf.

12.   This paper was not cited in the paleoclimate chapter (chapter 5) of the
AR5 's Working Group I report despite the fact that the lead author of the
paper, Dr. Dabang Jiang is an expert reviewer for AR5 report: D. Jiang,
X. Lang, Z. T. T. Wang, "Considerable model-data mismatch in tempera-
ture over China during the mid-Holocene: Results of PMIP simulations,"
Journal of Climate Vol. 25 (2012), 4135-4153.

13.   Ibid. See Figure 2d and Table 3.

14.   S. Mekaoui and S Dewitte, "Total Solar Irradiance measurement and

modelling during cycle 23," *Solar Physics*, Vol. 247 No. 1 (2008), 203-216; N. Scafetta and R. C. Willson, "ACRIM total solar irradiance composite validation versus TSI proxy models," *Astrophysics and Space Science*, Vol. 50 No. 2 (2014), 421-442.

15. A. BenMoussa, S. Gissot, U. Schühle, G. Del Zanna, F. Auchère, S. Mekaoui, A.R. Jones, D. Walton, C.J. Eyles, G. Thuillier, D. Seaton, I. E. Dammasch, G. Cessateur, M. Meftah, V. Andretta, D. Berghmans, D. Bewsher, D. Bolsée, L. Bradley, D. S. Brown, P. C. Chamberlin, S. Dewitte, L. V. Didkovsky, M. Dominique, F. G. Eparvier, T. Foujols, D. Gillotay, B. Giordanengo, J. P. Halain, R. A. Hock, A. Irbah, C. Jeppesen, D. L. Judge, M. Kretzschmar, D. R. McMullin, B. Nicula, W. Schmutz, G. Ucker, S. Wieman, D. Woodraska, and T. N. Woods, "On-orbit degradation of solar instruments," *Solar Physics*, Vol. 288 (2013), 389-434.

16. Scafetta and Willson, "ACRIM total solar irradiance composite validation versus TSI proxy models," as above.

17. For example, in a 2014 paper, Meftah et al. reported that the SOVAP (SOlar VAriability Picard) radiometer onboard Picard spacecraft measured the total solar irradiance for the summer of 2010 to be 1362.1 Wm^{-2} with an uncertainty of ± 2.4 Wm^{-2}: see M. Meftah, S. Dewitte, A. Irbah, A. Chevalier, C. Conscience, D. Crommelynck, E. Janssen, S. Mekaoui, "SOVAP/*Picard*, a spaceborne radiometer to measure the Total Solar Irradiance," *Solar Physics*, Vol. 289 No. 5 (2014), 1885-1899.

18. As another example of contemporary 'pal-review' practices in science, a recent paper has announced that simulations of Earth climate using two absolute values of TSI of 1367.0 and 1361.3 Wm^{-2} only led to 'understandable, repeatable, small' differences—see D. H. Rind, J. L. Lean and J. Jonas, "The impact of different solar irradiance values on current climate model simulations," *Journal of Climate*, Vol. 27 (2014), 1100-1120. These authors achieved their miraculous result by using a 'compensation' approach that involved tuning relative humidity threshold (from 82.7 per cent to 83.1 per cent) for forming clouds and global cloud cover (from

58.3 per cent to 57.8 per cent) at will. Such a convenient conclusion by Rind et al. is clearly contradicted by simulations of unforced variations in global temperature in M.-C. Liang, L.C. Lin, K.K Tung, Y. L. Yung and S. Sun, "Impact of climate drift on twenty-first-century projection in a coupled atmospheric-ocean general circulation model," *Journal of the Atmospheric Sciences*, Vol. 70 (2013), 3321-3327. Using similar Goddard Institute for Space Studies climate models, the latter study shows changes by as large as 1°C over 4000 years. In order to fully arrest such artificial model tuning exercises, consider the idealised numerical experiments performed using other atmospheric general circulation models. In J. Barsugli, S.I. Shin and P. D. Sardesmukh, "Tropical climate regimes and global climate sensitivity in a simple setting," *Journal of Atmospheric Sciences*, Vol. 62 (2005), 1226-1240, if a solar constant value of 1360 Wm^{-2} is prescribed, an overheated global mean surface temperature of 38°C, instead of the observed 14°C to 15°C, is produced. In order to examine how atmospheric deep moist convection interacting with large-scale flow on the equal global temperature footings, B. P. Kirtman and E. K. Schneider, "A spontaneously generated tropical atmospheric general circulation," *Journal of Atmospheric Sciences*, Vol. 57 (2000), 2080-2093 have to lower the baseline solar constant value of 1365 Wm^{-2} to 990 Wm^{-2}—a reduction by 375 Wm^{-2}—in order to produce 'an energetically consistent atmospheric general circulation'.

19.   It should be noted that we leave the original units for solar irradiance as it is here to be compared, albeit indirectly, with the so-called climatic forcing units of energy flux in the climate system without any conversion by assuming a certain value of planetary albedo and geometrical factors (see earlier discussion on the choice of solar zenith angle). The reason is that we wish to remind all scientists that the key physical processes and mechanisms leading to our observed cloud fields and planetary albedo are still not understood (see, for example, F. Hoyle, "The Great Greenhouse Controversy" in *The Global Warming Debate: The Report of the European*

Science and Environment Forum (Dorset: Bourne Press Limited, 1996) 180) and that resolving this one particular issue will certainly lead to a much improved state of knowledge.

20. Soon and Legates, "Solar irradiance modulation," as above.

21. R. S. Lindzen, "Climate dynamics and global change," *Annual Review of Fluid Mechanics*, Vol. 26 (1994), 353-378.

22. Steinhilber et al., "Total solar irradiance during the Holocene," as above; Steinhilber et al., "9,400 years of cosmic radiation and solar activity," as above.

23. Dr. Fröhlich has since retired and Dr. Steinhilber has left academia, and so these authors are no longer available to address concerns about the weaknesses of their studies. As a consequence, important errors may not be corrected or will take too long to be corrected if I remain silent and do not raise the issue.

24. A total of four points were shown in Figure 4c of Fröhlich ("Evidence of a long-term trend in total solar irradiance," as above) but two of the points in the figure should not have been included: one is from an extrapolation back to 1976/1977 and another one is for the most recent minimum of sunspot Cycle 23/24 at around 2008 during the time of publication, before the solar activity minimum can actually be decided.

25. The 9300-year reconstruction of total solar irradiance by Steinhilber et al. ("9,400 years of cosmic radiation and solar activity," as above; "Total solar irradiance," as above) is featured in the report of Working Group I in the IPCC's *Fifth Assessment Report*: see T. F. Stocker, D. Qin, G. K. Plattner, M. M. B. Tignor, S. K. Allen, J. Boschung, A. Nauels, Y. Xia, V. Bex, P. M. Midgley, eds., *Climate Change 2013: The Physical Science Basis—Working Group I Contribution to the Fifth Assessment Report of the Intergovernmental Panel on Climate Change* (Cambridge: Cambridge University Press, 2013), 390.

26. G. A. Schmidt, J. H. Jungclaus, C. M. Ammann, E. Bard, P. Braconnot, T. J. Crowley, G. Delaygue, F. Joos, N. A. Krivova, R. Muscheler, B. L.

Otto-Bliesner, J. Pongratz, D. T. Shindell, S. K. Solanki, F. Steinhilber and L. E. A. Vieira, "Climate forcing reconstructions for use in PMIP simulations of the last millennium (v1.0)," *Geoscientific Model Development*, Vol. 4 (2011), 33-45.

27.   Fontenla et al., "High-resolution solar spectral irradiance," as above. Fontenla et al. in a 2014 paper have recently improved and updated their solar spectral irradiance in the far (122-200 nm) and extreme (10-121 nm) ultraviolet spectral regions. J. M. Fontenla, E. Landi, M. Snow and T. Woods, "Far and extreme-UV solar spectral irradiance and radiance from simplified atmospheric physical models," *Solar Physics*, Vol. 289 (2014), 515-544.

28.   W. Livingston and M. J. Penn, "Are sunspots different during this solar minimum?" *Eos, Transactions, American Geophysical Union*, Vol. 90 (2009), 257, 264; M. J. Penn and W. Livingston, "Long-term evolution of sunspot magnetic fields" in *The Physics of the Sun and Star Spots* (International Astronomical Union, IAU Symposium S273, 2011), 126-133.

29.   Livingston and Penn, "Are sunspots different during this solar minimum?", as above.

30.   The decaying umbral magnetic field strength tendency appears to still hold for data updated through February 2014 (Dr. Livingston, email message to author, 7 March 2014).

31.   The goal of *Kepler's* mission is not only to study Sun-like stars but also to discover Earth-like planets. See E. A. Petigura, A. W. Howard, G. W. Marcy, "The Prevalence of Earth-size planets orbiting Sun-stars," *Proceedings of the National Academy of Sciences*, Vol. 110 (2013), 19273-19278. Petigura *et al.* recently quantified, based on statistics accumulated from the *Kepler's* mission, that 11±4 per cent of the Sun-like stars may harbor an Earth-sized planet receiving between one and four times the stellar irradiation intensity as that on Earth.

32.   G. Basri, L. M. Walkowicz and A. Reiners, "Comparison of *KEPLER*

photometric variability with the Sun on different timescales," *The Astrophysical Journal*, Vol. 769 No. 1 (2013), 37. See also, T.J. Henry, D.R Soderblom, R.A. Donahue, S.L Baliunas, "A survey of Ca II H and K chromospheric emission in southern solar-type stars." *The Astronomical Journal*, vol. 111(1996), 439-465.

33.   The term 'para-science' alludes to a kind of intellectual parasitism. It may be defined as going through the motions of science without actually doing science: see C. Essex, "Book review— *The Climate Caper* by Garth Paltridge," *Energy & Environment*, Vol. 21 (2009), 1407-1410.

34.   D.R. Legates, W.Soon, W.M. Briggs, C.Monckton, "Climate Consensus and 'Misinformation': a Rejoinder to Agnotlogy, Scienctific Consensus, and the Teaching and Learning of Climate Change", *Science and Education*, August 2013, accessed 19 November 2014, http://link. springer.com/article/10.1007/s11191-013-9647-9

Chapter 5
The scientific context

1.   I thank Craig Idso, Fred Singer and Alan Moran for their constructive critical comments on the draft manuscript. The material in Tables 2 and 3 is extracted from the Summary for Policymakers of the IPCC and NIPCC reports, respectively (references below), and Figure 1 is reproduced from the following book: R. M. Carter, J. Spooner, W. Kininmonth, M. Feil, S. Franks and B. Leyland, *Taxing Air* (Melbourne: Kelpie Press, 2013).

2.   R. M. Carter, *Climate: the Counter Consensus* (London: Stacey International, 2010).

3.   P. Brohan, J. Kennedy, S. Tett, I. Harris and P. Jones, "Report on HadCRUT3 including error estimates," (Climate Research Unit, University of East Anglia), last modified 29 March 2005, accessed 14 July 2014, http://strata-sphere.com/blog/wp-content/uploads/hadcrut3_gmr_

defra_report_200503.pdf.

4. P. Brohan, J. J. Kennedy, I. Harris, S. F. B. Tett, and P. D. Jones, "Uncertainty estimates in regional and global observed temperature changes: a new dataset from 1850," *Journal of Geophysical Research* Vol. 111 (2006).

5. Compare D. T. Avery and S. F. Singer, *Unstoppable Global Warming: Every 1,500 Years* (Lanham: Rowman & Littlefield Publishers, 2008).

6. C. D. Idso and S. W. Idso, *The Many Benefits of Atmospheric CO_2 Enrichment* (Pueblo West: Enrichment Lake Publishing, 2011).

7. See Carter, *Climate*, as above, chapters 7-9.

8. J. Brignell, "A complete list of things caused by global warming," *NumberWatch*, last modified 3 March 2012. http://www.numberwatch.co.uk/warmlist.htm.

9. J. Spooner in Carter et al. *Taxing Air*, as above, 9.

10. S. Goldenburg, "IPCC chairman dismisses climate report spoiler campaign," *The Guardian*, September 2013, accessed 14 July 2014, http://www.theguardian.com/environment/2013/sep/19/ipcc-chairman-climate-report.

11. J. Cribb, "A Matter of Trust," *Australasian Science*, Vol. 23 No. 7 (2002), 38.

12. IPCC, *Climate Change 2013—The Physical Science Basis. Working Group I Contribution to the Fifth Assessment Report of the Intergovernmental Panel on Climate Change* (Cambridge: Cambridge University Press, 2013); *IPCC, Climate Change 2014: Impacts, Adaptations and Vulnerability. Working Group II Contribution to the Fifth Assessment Report of the Intergovernmental Panel on Climate Change* (Cambridge: Cambridge University Press, 2014.)

13. NIPCC, C. D. Idso, R. M. Carter and F. W. Singer, eds, *Climate Change Reconsidered II—Physical Science.* (Chicago: Heartland Institute, 2013); NIPCC, C. D. Idso, S. W. Idso, R. M. Carter and F. W. Singer, eds, *Climate Change Reconsidered II—Biological Impacts.* (Chicago: Heartland Institute, 2014).

14. See D. Laframboise, *The Delinquent Teenager Who Was Mistaken For the World's Top Climate Expert,* (CreateSpace Independent Publishing Platform, 2011); D. Laframboise, *Into the dustbin: Rajendra Pachauri, the climate*

report and the Nobel Peace Prize (CreateSpace Independent Publishing Platform, 2013).

15.   cf. Carter et al. *Taxing Air*, as above, 228.

Chapter 6
Forecasting rain

1.   J. A. Banobi, T. A. Branch and R. Hilborn, "Do rebuttals affect future science?" *Ecosphere*, Vol. 2 No. 3 (2011).

2.   T. S. Kuhn, *The Structure of Scientific Revolutions* (Chicago: University of Chicago Press, 1962).

3.   J. Vaze, A. Davidson, J. Teng and G. Podger, "Impact of climate change on water availability in the Macquarie-Castlereagh River Basin in Australia," *Hydrological Processes*, Vol. 25 No. 16 (2011), 2597-2612.

4.   J. Abbot and J. Marohasy, "Input selection and optimization for monthly rainfall forecasting in Queensland, Australia, using artificial neural networks," *Atmospheric Research*, Vol. 138 (2014), 166-178.

5.   D. B. Irving, P. Whetton, A. F. Moise, "Climate Projections in Australia: a first glance at CMIP5," *Australian Meteorological and Oceanographic Journal*, Vol. 62 No. 4 (2012), 211-225.

6.   R. M. Carter, *Climate: The Counter-Consensus—A Palaeoclimatologist Speaks* (London: Stacey International, 2010).

7.   H. Halide and P. Ridd, "Complicated ENSO models do not significantly outperform very simple ENSO models," *International Journal of Climatology*, Vol. 28, No. 2 (2008), 219-233.

8.   A. Wu, W. W. Hsieh and B. Tang, "Neural network forecasts of the tropical Pacific sea surface temperatures," *Neural Networks*, Vol 19, No. 2 (2006), 145-154

9.   A. Kellow, *Science and Public Policy: The Virtuous Corruption of Virtual Environmental Science* (Bodmin: Edward Elgar Pub, 2007).

Chapter 7
Cool it: an essay on climate change

1. This essay is based on the text of a speech given in March 2014 to the Institute for Sustainable Energy and the Environment at the University of Bath.

2. J. Curry. *Statement to the Committee on Environment and Public works of the United States Senate.* Hearing on "Review of the President's Climate Action Plan", 16 January 2014, p7, accessed 19 November 2014, http://www.epw.senate.gov/public/index.cfm?FuseAction=Files.View&FileStore_id=07472bb4-3eeb-42da-a49d-964165860275

3. D. Campbell-Lendrum, D.D. Chadee, Y. Honda, Q. Liu, J.M.Olwoch, B. Revich, R. Sauerborn, "Human Health: Impacts, Adaptation, and Co-Benefits" in *IPCC Climate Change 2014: Impacts, Adaptation, and Vulnerability. Working Group II Contribution to Fifth Assessment Report of the Intergovernmental Panel on Climate Change*, eds. K.R. Smith, A.Woodward (Cambridge: Canbridge University Press, 2014) p713

Chapter 8
Costing climate change

1. M. Beenstock, Y. Reingewertz and N Paldor, "Polynomial cointegration tests of anthropogenic impact on global warming," *Earth System Dynamics*, Vol. 3 (2012): 173-188, accessed 10 July 2014, http://www.earth-syst-dynam.net/3/173/2012/esd-3-173-2012.pdf.

2. S. Michaels, "U2 criticised for world tour carbon footprint", *The Guardian*, 10 July, 2009, accessed 19 November 2014, http://www.theguardian.com/music/2009/jul/10/u2-world-tour-carbon-footprint

3. IPCC, "Summary for policy makers," *Climate Change 2014: Impacts, Adaptation, and Vulnerability. Part A: Global and Sectoral Aspects. Contribution of Working Group II to the Fifth Assessment Report of the*

Intergovernmental Panel on Climate Change (Cambridge, Cambridge University Press, 2014), 12, accessed 10 July 2014, http://ipcc-wg2.gov/AR5/images/uploads/IPCC_WG2AR5_SPM_Approved.pdf.

4. UK: N. Stern, *The Economics of Climate Change: The Stern Review* (London: HM Treasury, 2006), i-xxvii, accessed 10 July 2014 at http://webarchive.nationalarchives.gov.uk/20130129110402/http://www.hm-treasury.gov.uk/d/Executive_Summary.pdf. Australia: R. Garnaut, *The Garnaut Climate Change Review: Final Report* (Melbourne: Cambridge University Press, 2008), 254, accessed 10 July 2014, http://www.rossgarnaut.com.au/Documents/GCCR%20final%20report%20pdf/Garnaut_Chapter11.pdf.

5. Commonwealth of Australia 2011, "Strong Growth, Low Pollution", accessed 19 November 2014, http://cache.treasury.gov.au/treasury/carbonpricemodelling/content/report/downloads/Modelling_Report_Consolidated.pdf

6. Garnaut, *The Garnaut Climate Change Review*, as above, 249.

7. Australia's Treasury Secretary, Martin Parkinson, formerly the Secretary of the Department of Climate Change, seemed unaware of this when, in April 2014, he, 'told an audience in Washington it appeared inevitable that Australia would have to resettle climate change refugees in the coming decades.' See N. O'malley, "Treasury chief Martin Parkinson says climate refugees inevitable," *Sydney Morning Herald*, 13 April 2014, accessed 10 July 2014, http://www.smh.com.au/federal-politics/political-news/treasury-chief-martin-parkinson-says-climate-refugees-inevitable-20140412-36k47.html.

8. IPCC, "Summary for Policymakers," as above, 19.

9. K. Rehdanz and D. Maddison, "The Impact of Climate on Happiness and Life-Satisfaction," *Ecological Economics*, Vol. 70 No. 12 (2011), 2437-2445.

10. Nordhaus, *A Question of Balance*, Yale University Press (2008).

11. F. Bosello, F. Eboli and R. Pierfederici, "Assessing the Economic Impacts

of Climate Change," *Review of Environment, Energy and Economics (Re3)* (2012), accessed 10 July 2014, http://www.feem.it/userfiles/ attach/20126221213564Re3-BoselloEboliPierfederici-20121002.pdf.

12. R. Roson and D. Van der Mensbrugghe, "Climate change and economic growth: impacts and interactions," *International Journal of Sustainable Economy*, Vol. 4 No. 3 (2012), 270-285.

13. "Climate change effects already 'widespread and consequential' says IPCC co-chair," Australian Broadcasting Corporation, accessed July 10, 2014, http://www.abc.net.au/7.30/content/2014/s3975401.htm.

14. J. Wilcox and D. Makowski, "A meta-analysis of the predicted effects of climate change on wheat yields using simulation studies," *Field Crops Research* Vol. 156 (2014), 180-190.

15. IPCC, "Summary for Policymakers," as above.

16. Stern, *The Economics of Climate Change*, as above.

17. Ibid, 239.

18. Nongovernmental International Panel on Climate Change (NIPCC), *Climate Change Reconsidered II: Human Welfare, Energy, and Policies* (Chicago: The Heartland Institute, 2014), 53.

19. E. Lanzi, D. Mullaly, J. Chateau and R. Dellink, "Addressing Competitiveness and Carbon Leakage Impacts Arising from Multiple Carbon Markets: A Modelling Assessment," *OECD Environment Working Papers* No. 58 (OECD Publishing: 2013) 25, accessed 10 July 2014 at http://www.oecd-ilibrary.org/docserver/download/5k40ggjj7z8v.pdf?ex pires=1404966169&id=id&accname=guest&checksum=3DBA0CF291 0F472EBC29F4B467142680. Measures, like renewable requirements, standards etc. have not been incorporated within the estimates, hence the true tax is greater than that estimated in the table.

20. Developed countries' carbon dioxide emissions from consumption are actually broadly similar. See S. J. Davis and K. Caldeira, "Consumption-based accounting of CO_2 emissions," *Proceedings of the National Academy of Sciences of the United States of America* Vol. 107 No. 12, accessed 10 July

2014 www.pnas.org/cgi/doi/10.1073/pnas.0906974107.

21.   A. Moran."Aussies won't pay for climate schemes", April 2014, accessed 19 November 2014 http://www.ipa.org.au/portal/uploads/Media_ Release-Poll-Aussies_wont_pay_for_climate_schemes.pdf

22.   See "Australia's Electricity (Slowly) Getting Greener," *Energy Matters*, accessed 10 July 2014, http://www.energymatters.com.au/renewable-news/em988/.

Chapter 9
Experts as ideologues

1.   C. P. Snow, *The Two Cultures* (New York: Cambridge University Press, 1964).

2.   C. Lucas quoted in the *Daily Mail*: T.Clark, "Air travel 'as bad as stabbing person in the street', says MEP", *Daily Mail*, 20 April 2009, accessed 19 November 2014, http://www.dailymail.co.uk/travel/article-1169862/Air-travel-bad-stabbing-person-street says-MEP.html#ixzz3JUxRzPIK

3.   M. Gerson "Americans' aversion to science carries a high price", *The Washington Post,* 12 May 2014, accessed 19 November 2014, http://www.washingtonpost.com/opinions/michael-gerson-americans-aversion-to-science-carries-a-high-price/2014/05/12/7800318e-d9fe-11e3-bda1-9b46b2066796_story.html

4.   J. Ashton, "The BBC betrayed its values by giving Professor Carter this climate platform," *The Guardian,* 2 October 2013, accessed 19 November 2014, http://www.theguardian.com/commentisfree/2013/oct/01/bbc-betrayed-values-carter-scorn-ipcc

5.   "Lennard Bengtsson Resigns: GWPF voices shock and concern at the extent of intolerance within the climate science community," *The Global Warming Policy Foundation*, 14 May 2014, accessed 19 November 2014, http://www.thegwpf.org/lennart-bengtsson-resigns-gwpf-voices-shock-and-concern-at-the-extent-of-intolerance-within-the-climate-science-community/

6.   J. Schnell, "Our Fragile Earth," *Discover,* October 1989, p45-48.

Chapter10
Uncertainty, scepticism and the climate issue

1.   This chapter is based on the essay "Climate Change's Inherent Uncertainties" first published in the Jan-Feb 2014 edition of *Quadrant Magazine*. As well as new material, this chapter also contains several paragraphs from the article "Science held hostage in climate debate," published in the *Australian Financial Review*, on June 22, 2012.

Chapter 11
The trillion dollar guess based on a zombie theory

1.   See Climate Research Board, *Carbon Dioxide and Climate: A Scientific Assessment: Report of an Ad Hoc Study Group on Carbon Dioxide and Climate* (Woods Hole: National Academy of Sciences, 1979), accessed 14 July 2014, http://web.atmos.ucla.edu/~brianpm/download/charney_report.pdf.

2.   See A. Kossoy and P. Guigon, *State and Trends of the Carbon Market 2012* (Washington, D. C.: World Bank Institute, 2012), accessed 14 July 2014, http://siteresources.worldbank.org/INTCARBONFINANCE/Resources/State_and_Trends_2012_Web_Optimized_19035_Cvr&Txt_LR.pdf.

3.   K. Jones, "Bank of America Pledges $50 Billion to Combat Climate Change," *Capital.gr.*, 11 June 2012, accessed 14 July 2014, http://english.capital.gr/News.asp?id=1525641.

4.   See B. Buchner, M. Herve-Mignucci, C. Trabacchi, J. Wilkinson, M. Stadelmann, R. Boyd, F. Mazza, A. Falconer and V. Micale, *The Global Landscape of Climate Finance 2013* (San Francisco: Climate Policy Initiative, 2013).

5.   "Climate Gets 20 Percent of Seven-Year European Budget," *Environment News Service*, 8 February 2013, accessed 14 July 2014, http://ens-newswire.com/2013/02/08/

climate-gets-20-percent-of-seven-year-european-budget/.

6. Note 'rising humidity' here means rising *specific* humidity—warm air holds more water vapor. In this case *relative* humidity stays the same.

7. B. J. Soden, and I. M. Held, "An assessment of climate feedbacks in coupled ocean-atmosphere models," *Journal of Climate*, Vol. 19 No. 14 (2006), 3354-3360.

8. S. Sherwood, E. R. Kursinski and W. G. Read, "A Distribution Law for Free-Tropospheric Relative Humidity," *Journal of Climate*, Vol. 19 No. 24 (2006), 6267-6277.

9. Climate Research Board, "Carbon Dioxide and Climate," as above, 8.

10. G. Paltridge, A. Arking and M. Pook, "Trends in middle- and upper-level tropospheric humidity from NCEP reanalysis data," *Theoretical and Applied Climatology*, Vol. 98 No. 3-4 (2009), 351-359.

11. See D. H. Douglass, J.R. Christy, B.D. Pearson and S.F. Singer, "A comparison of tropical temperature trends with model predictions," *International Journal of Climatology*, Vol. 28 No. 13 (2007) 1693-1701; R. R. McKitrick, S. McIntyre and C. Herman "Panel and Multivariate Methods for Tests of Trend Equivalence in Climate Data Series," *Atmospheric Science Letters*, Vol. 11 No. 4 (2010), 270-277; J. R. Christy, B. Herman, R. Pielke, P. Klotzbach, R.T. McNide, J.J. Hnilo, R.W. Spencer, T. Chase and D. Douglass, "What Do Observational Datasets Say about Modeled Tropospheric Temperature Trends since 1979?," *Remote Sensing*, Vol. 2 No. 9 (2010), 2148-2169; Q. Fu, S. Manabe and C. Johanson, "On the warming in the tropical upper troposphere: Models vs observations," *Geophysical Research Letters*, Vol. 38, No. 15 (2011).

12. S. Huang, H. N. Pollack and P. Y. Shen, "Late Quaternary temperature changes seen in world-wide continental heat flow measurements," *Geophysical Research Letters*, Vol. 24 No.15 (1997), 1947-1950; S. P. Huang, H. N. Pollack and P.-Y. Shen, "A late Quaternary climate reconstruction based on borehole heat flux data, borehole temperature data, and the instrumental record," *Geophysical Research Letters*,

Vol. 35 (2008), accessed 14 July 2014, http://www.earth.lsa.umich. edu/~shaopeng/2008GL034187.pdf;—for a discussion, see J. Nova, "The message from boreholes," *Jo Nova: Skeptical Science for dissident thinkers*, accessed 14 July 2014, http://joannenova.com.au/2012/11/ the-message-from-boreholes/.

13.   B. Christiansen and F. C. Ljungqvist, "The extra-tropical Northern Hemisphere temperature in the last two millennia: reconstructions of low-frequency variability," *Climate of the Past*, Vol. 8 No. 2 (2012), 765-786—see discussion at S. McIntyre, "Kinnard and the D'Arrigo-Wilson Chronologies," *Climate Audit*, 3 December 2011, accessed 14 July 2014, http://climateaudit.org/2011/12/03/kinnard-and-the-darrigo-wilson-chronologies/, which notes a lack of complete archives and code; see also F. C. Ljungqvist, P. J. Krusic, G. Brattström and H. S. Sundqvist, "Northern Hemisphere temperature patterns in the last 12 centuries," *Climate of the Past*, Vol. 8, No. 1 (2012), 227-249, and discussion at "A Data-Rich 1200-Year Temperature History of the Northern Hemisphere," *CO2 Science,* accessed 14 July 2014, http://www.co2science.org/articles/ V15/N28/C2.php.

14.   P. L. Munday, A. J. Cheal, D. L. Dixson, J. L. Rummer and K. E. Fabricius, "Behavioural impairment in reef fishes caused by ocean acidification at CO2 seeps," *Nature Climate Change*, Vol. 4 (2014), 487-492.

15.   H. von Storch, A. Barkhordarian, K. Hasselmann and E. Zorita, "Can climate models explain the recent stagnation in global warming?," *Academia. edu*, accessed 14 July 2014, http://www.academia.edu/4210419/ Can_climate_models_explain_the_recent_stagnation_in_global_warming.

16.   R. R. McKitrick, "Robust Measurement of the Duration of the Global Warming Hiatus," [DRAFT] accessed 3 May 2014.

17.   See *The Hockey Schtick*, accessed 14 July 2014, http://hockeyschtick. blogspot.com.au/.

18.   See, for example, University Corporation for Atmospheric Research, "Slowdown in Tropical Pacific Flow Pinned on Climate Change," 3 May

2006, accessed 14 July 2014, http://www.ucar.edu/news/releases/2006/walker.shtml.

19. As suggest M. H. England, S. McGregor, P. Spence, G. A. Meehl, A. Timmermann, W. Cai, A. S. Gupta, M. J. McPhaden, A. Purich and A. Santosos, "Recent intensification of wind-driven circulation in the Pacific and the ongoing warming hiatus," *Nature Climate Change*, Vol 4. (2014), 222-227.

20. See R. R. Neely, III, O. B. Toon, S. Solomon, J.-P. Vernier, C. Alvarez, J. M. English, K. H. Rosenlof, M. J. Mills, C. G. Bardeen, J. S. Daniel and J. P. Thayer "Recent anthropogenic increases in SO2 from Asia have minimal impact on stratospheric aerosol," *Geophysical Research Letters*, Vol. 40 No. 5 (2013), 999-1004.

21. J. P. Vernier, L. W. Thomason, J. P. Pommereau, A. Bourassa, J. Pelon, A. Garnier, A. Hauchecorne, L. Blanot, C. Trepte, D. Degenstein and F. Vargas "Major influence of tropical volcanic eruptions on the stratospheric aerosol layer during the last decade," *Geophysical Research Letters*, Vol. 38 No. 12 (2011), accessed 14 July 2014, http://onlinelibrary.wiley.com/doi/10.1029/2011GL047563/abstract.

22. See X. Chen, Y. Feng and N. E. Huang, "Global Sea Level Trend during 1993-2012," *Global and Planetary Change*, Vol. 112 (2013), 26-32; A. Cazenave, H. Dieng, B. Meyssignac, K. von Schuckmann, B. Decharme and E. Berthier, "The rate of sea-level rise," *Nature Climate Change*, Vol. 4 (2014), 358-361; N. A. Mörner, *Sea Level is Not Rising* (Science & Public Policy Institute: 2012).

23. S. Levitus, J. I. Antonov, T. P. Boyer, O. K. Baranova, H. E. Garcia, R. A. Locarnini, A. V. Michonov, J. R. Reagon, D. Seidov, E. S. Yarosh and M. M. Zweng, "World ocean heat content and thermosteric sea level change (0-2000 m), 1955-2010," *Geophysical Research Letters*, Vol. 39 No. 10 (2012), accessed 14 July 2014, http://onlinelibrary.wiley.com/doi/10.1029/2012GL051106/pdf.

24. Cazenave et al., "The rate of sea-level rise," as above.

25. See G. G. Anagnostopoulos, D. Koutsoyiannis, A. Christofides, A. Efstratiadis and N. Mamassis, "A comparison of local and aggregated climate model outputs with observed data," *Hydrological Sciences Journal*, Vol. 55 No.7 (2010), 1094-1110; D. Koutsoyiannis, A. Efstratiadis, N. Mamassis and A. Christofides, "On the credibility of climate predictions," *Hydrological Science Journal*, Vol. 53 No. 4 (2008), 671-684.

26. M. Previdi and L. M. Polvani, "Climate system response to strato-spheric ozone depletion and recovery," *Quarterly Journal of the Royal Meteorological Society*, forthcoming (2014), accessed July 14, 2014, http://onlinelibrary.wiley.com/doi/10.1002/qj.2330/abstract.

27. See Christy et al. "What Do Observational Datasets Say about Modeled Tropospheric Temperature Trends since 1979?," as above; Q. Fu, S. Manabe and C. Johanson, "On the warming in the tropical upper troposphere: Models vs observations," *Geophysical Research Letters*, Vol. 38 (2011); see also discussion at R. A. Pielke, "New Paper Illustrates Another Failure of the IPCC Multi-Decadal Global Model Predictions—'Of the Warming in the Tropical Upper Troposphere: Models Versus Observations' by Fu et Al 2011," *Climate Science: Roger Pielke Sr.*, 8 July 2011, accessed 14 July 2014, http://pielkeclimatesci.wordpress.com/2011/07/08/new-paper-illustrates-another-failure-of-the-ipcc-mullti-decadal-global-model-predictions-on-the-warming-in-the-tropical-upper-troposphere-models-versus-observations-by-fu-et-al-2011/.

28. Humidity: see Paltridge et al., "Trends in middle- and upper-level tropo-spheric humidity," as above; ; rainfall: Anagnostopoulos et al., "A compar-ison of local and aggregated climate model outputs," as above; drought: J. Sheffield, E. F. Wood and M. L. Roderick, "Little Change in global drought over the past 60 years," *Nature*, Vol. 491, No. 7424 (2012), 435-438; clouds: M. Al. Miller, V. P. Ghate and R. K. Zahn, "The Radiation Budget of the West African Sahel and Its Controls: A Perspective from Observations and Global Climate Models," *Journal of Climate*, Vol. 25 No. 17 (2012), 5976-5996.

29.   R. J. Allen and S. C. Sherwood, "Warming maximum in the tropical up-
      per troposphere deduced from thermal winds," *Nature geoscience*, Vol. 1,
      No. 6 (2008), 399-403.

30.   B. D. Santer, P. W. Thorne, L. Haimberger, K. E. Taylor, T. M. L. Wigley,
      J. R. Lanzante, S. Solomon, M. Free, P. K. Gleckler, P. D. Jones, T. R.
      Karl, S. A. Klein, C. Mears, D. Nychka, G. A. Schmidt, S. C. Sherwood
      and F. J. Wentz, "Consistency of modelled and observed temperature
      trends in the tropical troposphere," *International Journal of Climatology*,
      Vol. 28, No. 13 (2008), 1703-1722.

31.   IPCC, *IPCC WGI Fifth Assessment Report, Second Order Draft*,
      (Intergovernmental Panel on Climate Change, 2012), 218.

32.   R. A. Pielke, "Heat Storage Within the Earth System," *Bulletin of the
      American Meteorological Society*, Vol. 84, No 3 (2003), 331-335.

33.   NASA: "It's not the Heat, it's the Humidity," *The Earth Observatory*,
      accessed 14 July 2014, http://earthobservatory.nasa.gov/Features/
      WaterVapor/water_vapor3.php; IPCC: IPCC, *Contribution of Working
      Group I to the Fourth Assessment Report of the Intergovernmental Panel on
      Climate Change* (Cambridge: Cambridge University Press, 2007), 632.

34.   USGS Water Science School, "The Water Cycle: Water Storage in the
      Atmosphere," accessed 14 July 2014, http://water.usgs.gov/edu/watercy-
      cleatmosphere.html.

35.   J. Nova, *Climate Money: The Climate Industry: $79 billion so far—tril-
      lions to come* (Science and Public Policy Institute, 2009), accessed 14 July
      2014, http://scienceandpublicpolicy.org/images/stories/papers/originals/
      climate_money.pdf.

36.   See J. Nova, "Deutsche Bank—A Wunch of Bankers," *JoNova:
      Skeptical Science for dissident thinkers*, accessed 14 July 2014,
      http://joannenova.com.au/2010/09/deutsche-bank-a-wunch-of-
      bankers/. For the original paper, see M. E. Carr, R. F. Anderson,
      K. Brash, *Climate Change: Addressing the Major Skeptic Arguments*
      (Deutsche Bank Group, 2010), accessed 14 July 2014, www.

theravinaproject.org/DBCCAColumbiaSkepticPaper090710.
pdf; see also M. Fulton, "Balancing reason and risk," *Climate
Spectator*, 10 September 2010, accessed 14 July 2014, http://
www.businessspectator.com.au/article/2010/9/10/policy-politics/
balancing-reason-and-risk#comment-1634.

37.   D. Fischer and The Daily Climate, "'Dark Money' Funds Climate
Change Denial Effort," *Scientific American*, 23 December 2014, accessed 14 July 2014, http://www.scientificamerican.com/article/
dark-money-funds-climate-change-denial-effort/.

38.   R. Brulle, "Institutionalizing delay: foundation funding and the creation
of U.S. climate change counter-movement organizations," *Climatic
Change*, Vol. 122 No. 4 (2014), 681-694.

39.   See "The Koch Brothers: Funding $67,042,064 to groups denying climate
change science since 1997," *Greenpeace*, accessed 14 July 2014, http://
www.greenpeace.org/usa/en/campaigns/global-warming-and-energy/
polluterwatch/koch-industries/.

40.   Buchner et al., *The Global Landscape of Climate Finance 2013*, as above.

41.   See J. Cook, "There is no such thing as climate change denial," *The
Conversation*, 15 February 2013, accessed 14 July 2014, http://theconver-
sation.com/there-is-no-such-thing-as-climate-change-denial-11763.

42.   A. Seabrook, "Gore Takes Global Warming Message to Congress,"
National Public Radio, 21 March 2007, 14 July 2014, http://www.npr.
org/templates/story/story.php?storyId=9047642.

43.   See "Idiotic Comment of the Day: Martin Rees," *Australian Climate
Madness*, 21 February 2010, accessed 14 July 2014, http://australiancli-
matemadness.com/2010/02/21/idiotic-comment-of-the-day-martin-rees/
comment-page-1/.

44.   J. Hansen, A. Lacis, D. Rind, G. Russell, P. Stone, I. Fung, R. Ruedy
and J. Lerner, "Climate sensitivity: Analysis of feedback mechanisms,"
Climate Processes and Climate Sensitivity: AGU Geophysical Monograph 29
(American Geophysical Union, 1984), 130-163.

45.   See also S. Bony, R. Colman, V. M. Kattsov, R. P. Allan, C. S. Bretherton, J. L. Dufresne, A. Hall, S. Hallegatte, M. M. Holland, W. Ingram, D. A. Randall, D. J. Soden, G. Tselioudis and M. J. Webb, "How well do we understand and evaluate climate change feedback processes?" *Journal of Climate*, Vol. 19, No. 15 (2006), 3445-3482; IPCC, *Contribution of Working Groups I, II and III to the Fourth Assessment Report of the Intergovernmental Panel on Climate Change* (Geneva: IPCC, 2007), 631.

46.   Karl et al (2006), Climate Change Science Program (CCSP) 2006 Report, Chapter 1, 1958-1999. Synthesis and Assessment Report 1.1, 2006, CCSP, Chapter 1, p 25, based on Santer et al. 2000 [PDF].

47.   Karl et al (2006) Climate Change Science Program (CCSP) 2006 Report, Chapter 5, part E of Figure 5.7 in section 5.5 on page 116

48.   D. H. Douglass, J. R. Christy, B. D. Pearson and S. F. Singer, "A comparison of tropical temperature trends with model predictions," *International Journal of Climatology*, Vol. 28, No. 13 (2007), 1693-1701. See discussion at "Tropical Trends Stir Warming Debate," *World Climate Report*, 14 December 2007, accessed 14 July 2014, http://www.world-climatereport.com/index.php/2007/12/14/tropical-trends-stir-warming-debate/; "Tropical Atmospheric Temperature Trends: Simulations vs. Measurements," *CO2 Science*, accessed 14 July 2014, http://www.co2science.org/articles/V11/N13/C2.php.

49.   Santer et al., "Consistency of modelled and observed temperature trends in the tropical troposphere," as above.

50.   R. R. McKitrick et al., "Panel and Multivariate Methods," as above.

51.   Allen and Sherwood, "Warming maximum in the tropical upper troposphere," as above; see discussion by L. Motl, "Sherwood, Allen, and radiosondes," *the reference frame*, June 8, 2008, accessed 14 July 2014, http://motls.blogspot.com.au/2008/06/sherwood-allen-and-radiosondes.html.

52.   S. Sherwood, C. L. Meyer, R. J. Allen, "Robust Tropospheric Warming Revealed by Iteratively Homogenized Radiosonde Data," *Journal of Climate*, Vol. 21, No. 20 (2008) 5336.

53.   J. Cook, *The Scientific Guide to Global Warming Scepticism*, (skepticalscience.com, 2010), accessed 14 July 2014, http://www.skepticalscience.com/The-Scientific-Guide-to-Global-Warming-Skepticism.html.

54.   IPCC, *IPCC WGI Fifth Assessment Report*, as above, 116—in particular, see Figure 10.

55.   A. E. Dessler,and S. M. Davis, "Trends in tropospheric humidity from reanalysis systems," *Journal of Geophysical Research*, Vol. 115, No. D19 (2010), accessed 14 July 2014, http://onlinelibrary.wiley.com/doi/10.1029/2010JD014192/pdf.

56.   See R. van Dorland, "The (missing) tropical hot spot," *Climate Dialogue: Exploring different views on climate change*, accessed 14 July 2014, http://www.climatedialogue.org/the-missing-tropical-hot-spot/.

57.   S. Lewandowsky, K. Oberauer and C. E. Gignac, "NASA faked the moon landing—therefore (climate) science is a hoax: An anatomy of the motivated rejection of science," *Psychological Science*, accessed 14 July 2014, http://websites.psychology.uwa.edu.au/labs/cogscience/documents/LskyetalPsychScienceinPressClimateConspiracy.pdf.

58.   See abstract here: S. Lewandowsky, H. Cook, K. Oberauer and M. Marriott, "Recursive fury: conspiracist ideation in the blogosphere in response to research on conspiracist ideation," *Frontiers in Psychology*, Accessed 14 July 2014, http://journal.frontiersin.org/Journal/10.3389/fpsyg.2013.00073/full; for details of its retraction, see "Retraction of Recursive Fury: A Statement," *Frontiers*, accessed 14 July 2014, http://www.frontiersin.org/blog/Retraction_of_Recursive_Fury_A_Statement/812.

Chapter 12
Forecasting global climate change

1.   CSIRO, "Future climate scenarios for Australia," *State of the Climate—2014* (Canberra: CSIRO, 2014), accessed 29 April 2014,

http://www.csiro.au/Outcomes/Climate/Understanding/State-of-the-Climate-2014/Future-Climate-Scenarios-for-Australia.aspx.

2. "Future Climate Change", *United States Environmental Protection Agency*, accessed 29 April 2014, http://www.epa.gov/climatechange/science/future.html#Temperature.

3. N. Stern, quoted in *The Guardian*: C. Urquhart, "Flooding and storms in UK are clear signs of climate change, says Lord Stern", *The Guardian*, 13 February 2014, accessed 29 April 2014, http://www.theguardian.com/environment/2014/feb/13/flooding-storms-uk-climate-change-lord-stern.

4. We are grateful to Steve Goreham, David Legates, Craig Loehle, Alan Moran, and Willie Soon for their helpful suggestions on this chapter. Hester Green, Jen Kwok, Lynn Selhat, and Angela Sun provided useful suggestions on the writing. The responsibility for any errors or omissions remains with the authors.

5. As documented in D. Laframboise, *The delinquent teenager who was mistaken for the world's top climate expert* (Toronto: Ivy Avenue Press, 2011) and T. Ball, *The Deliberate Corruption of Climate Science* (Seattle: Stairway Press, 2014).

6. Research is summarised in J. S. Armstrong, *Long-range Forecasting* (New York: John Wiley, 1985), 138-144.

7. C. Cerf and V. Navasky, *The Experts Speak* (New York: Pantheon, 1984).

8. J. S. Armstrong, "The Seer-Sucker Theory: The Value of Experts in Forecasting," *Technology Review*, Vol. 82 No. 7 (1980), 16-24; P. E. Tetlock, *Expert political judgement: How Good Is It? How Can We Know?* (Princeton: Princeton University Press, 2005).

9. J. S. Armstrong ed., *Principles of Forecasting: A Handbook for Researchers and Practitioners* (New York: Springer, 2001.) In addition, the ForPrin.com web site provides a checklist of the forecasting principles and software that help users to determine which methods to use in a given situation.

10. D. A. Randall, R. A. Wood, S. Bony, R. Colman, T. Fichefet, J. Fyfe,

V. Kattsov, A. Pitman, J. Shukla, J. Srinivasan, R. J. Stouffer, A. Sumi and K. E. Taylor, "Climate Models and Their Evaluation," in *Climate Change 2007: The Physical Science Basis. Contribution of Working Group I to the Fourth Assessment Report of the Intergovernmental Panel on Climate Change*, ed. S. Solomon, D. Qin, M. Manning, Z. Chen, M. Marquis, K. B. Averyt, M. Tignor and H. L. Miller (Cambridge and New York: Cambridge University Press, 2007).

11.   The IPCC refused to provide the authors' email addresses.

12.   K. C. Green and J. S. Armstrong, "Global warming: forecasts by scientists versus scientific forecasts," *Energy and Environment*, Vol. 18 No. 7-8 (2007), 995-1019.

13.   J. S. Armstrong, K. C. Green and A. Graefe, "Golden Rule of Forecasting: Be conservative," *Journal of Business Research* (2014, forthcoming). A copy of the Golden Rule of Forecasting checklist of guidelines can be found at goldenruleofforecasting.com, as can a draft copy of the paper.

14.   P. J. Michaels, "Fighting fire with facts," *CATO Institute*, January 18, 1999, accessed 14 July 2014, http://www.cato.org/publications/commentary/fighting-fire-facts.

15.   See, for example, L. Festinger, H. W. Riecken and S. Schachter, *When Prophecy Fails: A social and psychological study of a modern group that predicted the destruction of the World* (Minneapolis: University of Minnesota Press: 1956) and C. D. Batson, "Rational processing or rationalization? The effect of disconfirmation on a stated religious belief," *Journal of Personality and Social Psychology*, Vol. 32 (1975), 176-184.

16.   See, for example, the following study: M. Mahoney, "Publication prejudices: An experimental study of confirmatory bias in the peer review system," *Cognitive Therapy and Research*, Vol. 1 No. 2 (1977), 161-175.

17.   For example, J. E. Hansen, "Defusing the global warming time bomb," *Scientific American* (March 2004), 68-77.

18.   See J. R. Christy, B. Herman, R. Pielke, P. Klotzback, R. T. McNider, J. J. Hnilo, R. W. Spencer, T. Chase and D. Douglass, "What do observational

datasets say about modelled tropospheric temperature trends since 1979?" *Remote Sensing*, Vol. 2 (2010), 2148-2169.

19. IPCC, J. T. Houghton, G. J. Jenkins and J. J. Ephraums, eds. *Climate change: The IPCC scientific assessment* (Cambridge: Cambridge University Press, 1990), xi; IPCC, J. T. Houghton, B. A. Callander and S. K. Varney , eds. *Climate change: The IPCC scientific assessment* (Cambridge: Cambridge University Press, 1992), 17.

20. Available from: "HadCRUT3 dataset," *Met Office Hadley Centre observations datasets*, January 2010, accessed 14 July 2014, http://hadobs.metoffice.com/hadcrut3/.

21. K. C. Green, J. S. Armstrong and W. Soon, "Validity of Climate Change Forecasting for Public Policy Decision Making," *International Journal of Forecasting*, Vol. 25 (2009), 826-832.

22. Note that there is evidence that the series tends to substantially overstate any warming trend due to weather station locations becoming increasingly surrounded by buildings, asphalt, and heat sources, and the deployment of more sensitive measurement instruments, together with unexplained adjustments to the temperature readings. See R. R. McKitrick and P. J. Michaels 2007, "Quantifying the influence of anthropogenic surface processes and inhomogeneities on gridded global climate data," *Journal Geophysical Research*, Vol. 112 (2007); R. R. McKitrick and N. Nierenberg, "Socioeconomic Patterns in Climate Data", *Journal of Economic and Social Measurement*, Vol. 35 No. 3-4 (2010), 149-175; R. R. McKitrick, "Encompassing tests of socioeconomic signals in surface climate data," *Climate Change*, Vol. 120 (2013), 95-107; A. Watts, E. Jones, S. McIntyre and J. R. Christy, "An area and distance weighted analysis of the impacts of station exposure on the U.S. Historical Climatology Network temperatures and temperature trends: pre-print draft discussion paper," 2012, accessed 14 July 2014, http://wattsupwiththat.files.wordpress.com/2012/07/watts-et-al_2012_discussion_paper_webrelease.pdf.

23. See, for example, G. J. Kukla and R. K. Matthews, "When will the pres-

ent interglacial end?" *Science*, Vol. 178 (1972), 190-191, and Watt's Earth Day speech.

24. For examples, see "Earth undergoing global COOLING since 2002," *climatedepot.com*, accessed 14 July 2014, http://www.climatedepot. com/2013/06/15/forget-the-temperature-plateau-earth-undergoing-glob-al-cooling-since-2002-climate-scientist-dr-judith-curry-attention-in-the-public-debate-seems-to-be-moving-away-from/, and George Kukla in an interview in Mari Krueger: M. Krueger, "An Unrepentant Prognosticator," *Geld Magazine*, accessed July 14, 2014 (2007), http://www.gelfmagazine. com/archives/an_unrepentant_prognosticator.php.

25. D. R. Legates, W. Soon, W. M. Briggs and C. Monckton, "Climate consensus and 'misinformation': A rejoinder to *Agnotology, Scientific Consensus, and the Teaching and Learning of Climate Change,*" *Science and Education* (August 2013) accessed 14 July 2014, http://link.springer.com/ article/10.1007/s11191-013-9647-9; see also J. Bast and R. Spencer, "The Myth of the Climate Change '97%'," *Wall Street Journal*, May 26, 2014, accessed 14 July 2014, http://online.wsj.com/news/articles/SB100014240 52702303480304579578462813553136?mod=rss_opinion_main.

26. See "Global Warming Petition Project," petitionproject.org.

27. T. C. Chamberlin, "The method of multiple working hypotheses," *Science*, Vol. 148 (1965), 754-759.

28. See a summary of cooling and warming alarms here: D. Gainor, "Fire and Ice," 3 November 2010, accessed 14 July 2014, http://www.mrc.org/ special-reports/fire-and-ice-0 .

29. For example, see F. M. Mattes, "Report of committee on glaciers," *Transactions, American Geophysical Union*, Vol.20, No. 4 (1939), 518-523.

30. K. C. Green, J. S. Armstron and W. Soon, "Validity of Climate Change Forecasting for Public Policy Decision Making," *International Journal of Forecasting*, Vol. 25 (2009), 826-832.

31. See R. S. Lindzen, "Climate Science: Is it currently designed to answer questions? *Euresis Journal*, Vol. 2 (2012), 161-192; W. Soon, S. Baliunas,

S. B. Idso, K. Y. Kondratyev and E. S. Posmentier, "Modeling climate effects of anthropogenic carbon dioxide emissions: unknowns and uncertainties," *Climate Research*, Vol. 18 (2001), 259-275.

32. Kukla and Matthews, "When will the present interglacial end?", as above, 190.

33. C. Loehle, "Climate change: detection and attribution of trends from long-term geologic data," *Econological modelling*, Vol. 171 (2004), 433-450.

34. See, for example, IPCC, "Summary for Policymakers," in *Climate Change 2013: The Physical Science Basis. Contribution of Working Group I to the Fifth Assessment Report of the Intergovernmental Panel on Climate Change,* T. F. Stocker, D. Qin, G. K. Plattner, M. Tignor, S. K. Allen, J. Boschung, A. Nauels, Y. Xia, V. Bex and P. M. Midgley, eds. (Cambridge: Cambridge University Press, 2013), 4.

35. C. Loehle and J. H. McCulloch, "Correction to: A 2000-year global temperature reconstruction based on non-tree ring proxies," *Energy & Environment*, Vol. 19, No. 1 (2008), 93-100.

36. See for example, W. Soon, S. Baliunas, C. Idso, S. Idso and D. R. Legates, "Reconstructing climatic and environmental changes of the past 1000 years: A reappraisal," *Energy & Environment*, Vol. 14 No. 2-3 (2003), 233-296; also see ongoing research reported by Medieval Warm Period Project: "Medieval Warm Period Project," *CO2 Science*, accessed 14 July 2014, http://www.co2science.org/data/mwp/mwpp.php.

37. S. Vavrus, W. F. Ruddiman and J. E. Kutzbach, "Climate model tests of the anthropogenic influence on greenhouse-induced climate change: the role of early human agriculture, industrialization, and vegetation feedbacks," *Quaternary Science Reviews*, Vol. 27 No. 13-14 (2008), 1410-1425.

38. Green et al., "Validity of Climate Change Forecasting for Public Policy Decision Making," as above.

39. K. C. Green and J. S. Armstrong, "The global warming alarm: Forecasts from the structured analogies method. Working paper," accessed 14 July

2014, http://www.kestencgreen.com/green&armstrong-agw-analogies.pdf.

40.   Green and Armstrong, "Global warming: forecasts by scientists versus scientific forecasts," as above.

41.   See, for example, K. Nikopoulos, A. Litsa, F. Petropoulos, V. Bougioukos and M. Khanmash, "Relative performance of methods for forecasting special events," *Journal of Business Research* (2014, forthcoming).

42.   Green and Armstrong, "The global warming alarm," as above.

43.   C. MacKay, *Extraordinary Popular Delusions and the Madness of Crowds* (New York: Three Rivers Press, 1841).

44.   Green and Armstrong, "The global warming alarm," as above. That site also includes a list of analogies that had been compiled by Julian Simon.

Chapter 13
The search for a global climate change treaty

1.   B. Ki-moon, "Remarks to the UNFCCC COP-15 closing plenary," *UN News Centre*, 19 December 2009, accessed 15 July 2014, http://www.un.org/apps/news/infocus/sgspeeches/search_full.asp?statID=686.

2.   "Copenhagen climate accord 'essential beginning': Ban," *Antaranews.com*, 20 December 2009, http://www.antaranews.com/en/news/1261284060/copenhagen-climate-accord-essential-beginning-ban.

3.   I. Traynor, "Global warming talks destined for disaster, says EU president," *The Guardian*, 4 December 2010.

4.   T. Stern, "A New Paradigm: Climate Change Negotiations in the Post-Copenhagen Era," 8 October 2010, accessed 15 July 2014, http://www.state.gov/e/oes/rls/remarks/2010/149429.htm.

5.   Ibid.

6.   T. Stern, interviewed in: "U.S. Preview of Climate Change Conference in South Africa," 23 November 2011, *IIP Digital—US Embassy*, accessed 14 July 2014, http://iipdigital.usembassy.gov/st/english/texttrans/2011/11/20111123142943su0.9419171.html#axzz37VpaadJX.

7. "U.S. Envoy Stern at COP-17 Conference in South Africa," *IIP Digital—US Embassy*, 8 December 2011, http://iipdigital.usembassy. gov/st/english/texttrans/2011/12/20111208170819su0.8296124. html?distid=ucs#axzz37VpaadJX.

8. H. Clinton, "U.S.-China Achievements Go Beyond Expo," *Global Times*, 21 May 2010, accessed 10 July 2014, http://www.state.gov/ secretary/20092013clinton/rm/2010/05/142073.htm.

9. J. Vidal, "Yvo de Boer: developing nations' suspicions slowing climate talks," *The Guardian*, 25 November 2010, accessed 15 July 2014, http://www.theguardian.com/environment/2010/nov/24/ yvo-de-boer-climate-change-cancun.

10. *Earth Negotiations Bulletin*, Vol. 12 No. 498 (13 December 2010), 16.

11. Ibid., 4.

12. C. Chauhan, "Jairam Ramesh admits deviating from India's stance," *Hindustan Times*, 9 December 2010, accessed 15 July 2014, http://www. hindustantimes.com/world-news/jairam-ramesh-admits-deviating-from-india-s-stance/article1-636250.aspx.

13. "Government faces Opposition fire for Cancun statement," *The Financial Express*, 10 December 2010, accessed 15 July 2014, http://www.financialexpress.com/news/govt-faces-oppn-fire-for-cancun-statement/723285.

14. "India calls for global energy hunt as demand set to soar," *Daily Times*, 1 November 2010, accessed 15 July 2014, http://archives.dailytimes.com. pk/business/02-Nov-2010/india-calls-for-global-energy-hunt.

15. E. Conway-Smith, "Climate change talks face challenges—South Africa: UN climate change summit opens in Durban," *GlobalPost*, 28 November 2011.

16. "Ban pleads for Kyoto in warning of climate deadlock," *Radio Netherlands Worldwide: Africa*, 6 December 2011, accessed 15 July 2014, http://www. rnw.nl/africa/bulletin/ban-pleads-kyoto-warning-climate-deadlock-1.

17. "Canada to drop Kyoto sooner than expected," *The Nation*, 7 December 2011, accessed 15 July 2014, http://www.nationmultimedia.com/break-

ingnews/Canada-drops-Kyoto-sooner-than-expected-30171734.html.

18. "What outcome for Durban climate talks?" *Times of Malta*, 27 November 2011

19. P. Clark and A. England, "Hard commitment sought for a phase 2 Kyoto treaty," *Financial Times*, 8 December 2011, accessed 15 July 2014, http://www.ft.com/cms/s/0/b473ba36-20f2-11e1-8a43-00144feabdc0.html#axzz37Vslrd8O.

20. DECC, "The Rt Hon Chris Huhne MP speech to the Durban COP17 Climate Conference Plenary," *Gov.UK*, 8 December 2011, accessed 15 July 2014, https://www.gov.uk/government/speeches/the-rt-hon-chris-huhne-mp-speech-to-the-durban-cop17-climate-conference-plenary.

21. A. Max, "In climate talks West would redefine rich and poor," *Associated Press*, 23 November 2011.

22. J. M. Broder, "US Climate Envoy Seems to Shift Stance on Timetable for New Talks," *New York Times*, 8 December 2011, accessed 15 July 2014, http://www.nytimes.com/2011/12/09/science/earth/us-climate-envoy-seems-to-shift-position-on-timetable-for-new-international-talks.html?_r=0.

23. Jayanthi Natarajan quoted in: "UN Climate Talks Head to Deadlock as EU Plan Hit by China, India", *Newsmax,* Saturday 10 December 2011, accessed 28 November 2014, http://www.Newsmax.com/Newsfront/ASIATOP-ASIATOPZ6-AUCURZ6-AUTOP/2011/12/10/id/420544/#ixzz3KKBTty6n.

24. United Nations Framework Convention on Climate Change, *Report of the Conference of the Parties on its seventeenth session, held in Durban from 28 November to 11 December 2011 Addendum Part Two: Action taken by the Conference of the Parties at its seventeenth session* (Durban: United Nations Framework Convention on Climate Change, 2012), 2, accessed 15 July 2014, http://unfccc.int/resource/docs/2011/cop17/eng/09a01.pdf.

25. A. Max, "Climate conference approves landmark deal," *The Washington Times*, 11 December 2011, accessed 15 July 2014, http://www.washingtontimes.com/news/2011/dec/10/climate-deal-approval-un-conference/?page=all.

26.   K. Ritter, "US envoy hits China's stand in UN climate talks," *The Washington Times*, 3 December 2012, accessed 15 July 2014, http://www.washingtontimes.com/news/2012/dec/3/ us-envoy-hits-chinas-stand-in-un-climate-talks/?page=all.

27.   "China Voice: Welcome to 'real world' of climate change," *China Daily*, 4 December 2012, accessed 15 July 2014, http://www.chinadaily.com.cn/ world/2012climate/2012-12/04/content_15984805.htm.

28.   *Earth Negotiations Bulletin*, Vol. 12 No. 567 (2012), 9.

29.   K. Ritter, "AP Interview: UN chief blames rich for warming," *Yahoo! News*, 5 December 2012, accessed 15 July 2014, http://news.yahoo.com/ ap-interview-un-chief-blames-rich-warming-085524684--finance.html.

30.   M. Casey, "UN climate talks go into overtime in Qatar," *USA Today*, 7 December 2012, accessed 15 July 2014, http://www.usatoday.com/story/ news/world/2012/12/07/un-climate-qatar/1753773/.

31.   *Earth Negotiations Bulletin*, Vol. 12 No.5 67 (2012), 28.

32.   M. Casey, "UN climate talks go into overtime in Qatar." *The Big Story*, 7 December 2012, accessed 17 July 2014, http://bigstory.ap.org/article/ money-focus-un-climate-talks-enter-last-day.

33.   "EU, US rule out climate funding pledges in Doha," *Gulf Times*, 5 December 2012, accessed 15 July 2014, http://www.gulf-times.com/qatar/178/ details/334679/eu,-us-rule-out-climate-funding-pledges-in-doha.

34.   P. Clark, "Doha talks agree to climate compensation," *Financial Times*, 8 December 2012, accessed 15 July 2014, http://www.ft.com/intl/cms/ s/0/53370cc6-413c-11e2-a517-00144feabdc0.html#axzz37Vslrd8O.

35.   *China Daily*, "Chinese official 'dismayed' at Japan's new emission target," *CCTV.com*, 15 November 2013, accessed 15 July 2014, http://english. cntv.cn/20131115/100345.shtml.

36.   S. Nicola, "Coal sector must 'change dramatically': UN climate chief," *Business Week*, 18 November 2013, accessed 15 July 2014, http://www.businessweek.com/news/2013-11-18/ coal-industry-must-change-dramatically-un-climate-chief-says.

37. K. Ritter, "Turmoil at climate talks as blame game heats up," *Daily Herald*, 20 November 2013, 15 July 2014, https://www.dailyherald.com/article/20131120/news/711209848/.

38. Mariette le Roux, "Troubled UN climate talks run into extra time," Mail & Guardian, 23 November 2013.

39. K. Ritter, "Compromise breaks deadlock at UN climate talks," *Daily Sun*, 23 November 2013, accessed 15 July 2014, http://www.daily-sun.com/details_yes_24-11-2013_Troubled-UN-climate-talks-run-into-extra-time_684_1_12_1_3.html.

40. "Historic agreement looms: UN talks approve climate pact principles," 23 November 2013, *The New Daily*, accessed 15 July 2013, http://thenew-daily.com.au/news/2013/11/23/un-talks-approve-climate-pact-principles/.

41. "Modest deal breaks deadlock at U.N. climate talks," *USA Today*, 23 November 2013, accessed 15 July 2014, http://www.usatoday.com/story/news/world/2013/11/23/modest-deal-breaks-deadlock-at-un-climate-talks/3686385/.

42. United Nations Framework Convention on Climate Change, *Report of the Conference of the Parties on its nineteenth session, held in Warsaw from 11 to 23 November 2013 Addendum Part two: Action taken by the Conference of the Parties at its nineteenth session*, (Warsaw: United Nations Framework Convention on Climate Change, 2014),

43. A. Max, "UN conference to deal with carbon reductions," *timesunion.com*, 27 November 2011, accessed 15 July 2014, http://www.timesunion.com/news/article/UN-conference-to-deal-with-carbon-reductions-2296275.php.

Chapter 14
The hockey stick: a retrospective

1. A. Montford, *The Hockey Stick Illusion* (London: Stacey International, 2010).

2. M. E. Mann, R. S. Bradley and M. K. Hughes, "Global-Scale

Temperature Patterns and Climate Forcing Over the Past Six Centuries," *Nature*, Vol. 392, No. 6678 (1998), 779-787; M. E. Mann, R. S. Bradley and M. K. Hughes, "Northern Hemisphere Temperatures During the Past Millennium: Inferences, Uncertainties, and Limitations," *Geophysical Research Letters*, Vol. 26 No. 6 (1999), 759-762.

3. S. McIntyre and R. McKitrick "Corrections to the Mann et al. (1998) Proxy Data Base and Northern Hemispheric Average Temperature Series," *Energy and Environment*, Vol. 14 No. 6 (2003), 751-771; S. McIntyre and R. McKitrick, "The M&M Critique of the MBH98 Northern Hemisphere Climate Index: Update and Implications," *Energy and Environment*, Vol. 16 No. 1 (2005), 69-99; S. McIntyre and R. McKitrick, "Hockey sticks, principal components, and spurious significance," *Geophysical Research Letters*, Vol. 32 No. 3 (2005), accessed 15 July 2014, http://climateaudit.files.wordpress.com/2009/12/mcintyre-grl-2005.pdf; S. McIntyre and R. McKitrick, "Reply to Comment by Von Storch and Zorita on 'Hockey sticks, principal components, and spurious significance,'" *Geophysical Research Letters*, Vol. 32 No. 20 (2005), accessed 15 July 2014, http://www.academia.edu/3318969/Reply_to_comment_by_von_Storch_and_Zorita_on_Hockey_sticks_principal_components_and_spurious_significance; S. McIntyre and R. McKitrick, "Reply to Comment by Huybers," *Geophysical Research Letters*, Vol. 32 (2005), accessed 15 July 2014, http://climateaudit.files.wordpress.com/2009/12/mcintyre-huybersreply.pdf.

4. See S. McIntyre and R. McKitrick, "Presentation to the National Academy of Sciences Expert Panel, 'Surface Temperature Reconstructions for the Past 1,000-2,000 Years'" (paper presented to National Academy of Sciences, Washington, D.C., 2 March 2006, accessed 15 July 2014, http://www.uoguelph.ca/~rmckitri/research/NAS.M&M.pdf.

5. S. McIntyre, "How do we 'know' that 1998 was the warmest year of the millennium?" (paper presented at Ohio State University, Columbus, Ohio, May 16, 2008), accessed 15 July 2014, http://climateaudit.files.

wordpress.com/2005/09/ohioshort.pdf.

6.   R. R. McKitrick, "What is the Hockey Stick Debate About? Presentation
     to the Asia Pacific Corporation Study Centre Meeting on 'Managing
     Climate Change—Practicalities and Realities in a post-Kyoto Future"
     (paper presented to Parliament House, Canberra, Australia, 4 April 2005),
     accessed 15 July 2014, http://www.rossmckitrick.com/paleoclimatehock-
     ey-stick.html; R. R. McKitrick, "The Mann et al. Northern Hemisphere
     'Hockey Stick' Climate Index: A Tale of Due Diligence," in *Shattered
     Consensus: The True State of Global Warming*, ed. P. Michaels (Lanham:
     Rowman & Littlefield, 2006).

7.   See S. McIntyre, *Climate Audit*, accessed 15 July 2014, http://climateaudit.org/.

8.   D. A. Graybill and S. B. Idso, "Detecting the aerial fertilization effect
     of atmospheric CO_2 enrichment in tree-rings chronologies," *Global
     Biogrochemical Cycles*, Vol. 7 No. 1 (1993), 81-95.

9.   McIntyre and McKitrick, "Hockey sticks, principal components, and
     spurious significance," as above.

10. On this, see Mann et al., "Global-Scale Temperature Patterns and Climate
     Forcing," as above, 786.

11. Ibid, 781-782—see Figure 3.

12. Ibid, 785.: "For comparison, correlation (r) and squared-correlation (r^2)
     statistics are also determined."

13. McIntyre and McKitrick, "Presentation to the National Academy of
     Sciences Expert Panels," as above, 11.

14. See detailed notes of the hearing: S. McIntyre, "Mann at the NAS Panel,"
     Climate Audit, 16 March 2006, accessed 15 July 2014, http://climateau-
     dit.org/2006/03/16/mann-at-the-nas-panel/.

15. See S. McIntyre and R. McKitrick, "Presentation to the National
     Academy of Sciences Expert Panel, 'Surface Temperature Reconstructions
     for the past 1,000-2,000 Years.': Supplementary Comments," 3 April
     2006, accessed 15 July 2014, http://www.rossmckitrick.com/up-
     loads/4/8/0/8/4808045/nas-followup-mm.pdf.

16.   G. R. North, F. Biondi, P. Bloomfield, J. R. Christy, K. M. Cuffey, R. E. Dickinson, E. R. M. Druffel, D. Nychka, B. Otto-Bliesner, N. Roberts, K. K. Turekian, J. M. Wallace and I. Kraucunas, *Surface Temperature Reconstructions for the last 2,000 years* (Washington, D.C.: The National Academy of Sciences, 2006).

17.   E. R. Wahl and C. M. Ammann, "Robustness of the Mann, Bradley, Hughest reconstruction of Northern Hemisphere surface temperatures: Examination of criticisms based on the nature and processing of proxy climate evidence," *Climatic Change*, Vol. 85 No. 1-2 (2007), 33-69.

18.   North et al., *Surface Temperature Reconstructions*, as above, 91.

19.   Ibid, 50, 107.

20.   S. McIntyre and R. McKitrick, "Corrections to the Mann et al. (1998)," as above.

21.   North et al., *Surface Temperature Reconstructions*, as above, 86-87.

22.   Ibid, 106.

23.   Ibid, 87.

24.   Ibid, 107.

25.   See M. E. Mann, E. Gille, R. S. Bradley, M. K. Hughes, J. T. Overpeck, F. T. Keimig and W. Gross, "Global temperature patterns in past centuries: An interactive presentation," *Earth Interactions*, Vol. 4 No. 4, 1-29 (2000), accessed 15 July 2014, http://www.ncdc.noaa.gov/paleo/ei/ eint_vol4_0004_1_29_2.pdf, 9: 'We have also verified that possible low-frequency bias due to non-climatic influences on dendroclimatic (tree-ring) indicators is not problematic in our temperature reconstructions': M. E. Mann, E. Gille, R. S. Bradley, M. K. Hughes, J. T. Overpeck, F. T. Keimig and W. Gross, "A Note on Possible Non-Climatic Tree-Ring Trend Bias," accessed 15 July 2014, http://www.ncdc.noaa.gov/paleo/ei/ei_no-dendro.html: "MBH98 found through statistical proxy network sensitivity estimates that skillful NH reconstructions were possible without using any dendroclimatic data ... Whether we use all data, exclude tree rings, or base a reconstruction only on tree rings, has no significant effect on the

form of the reconstruction for the period in question."

26. Mann et al., "Global temperature patterns in past centuries," as above.

27. Ibid.

28. The question and answer excerpt are in our NAS presentation—see: McIntyre and McKitrick, "Presentation to the National Academy of Sciences Expert Panel," as above, 21.

29. Ibid, 21-22.

30. Mann et al., "Northern Hemisphere Temperatures During the Past Millennium," as above, 760.

31. Mann et al., "Global temperature patterns in past centuries," as above.

32. Wahl and Ammann, "Robustness of the Mann, Bradley, Hughest recon-struction of Northern Hemisphere surface temperatures," as above.

33. McIntyre and McKitrick, "Presentation to the National Academy of Sciences Expert Panel, 'Surface Temperature Reconstructions for the past 1,000-2,000 Years.': Supplementary Comments," as above, 4.

Chapter 15
The IPCC and the peace prize

1. An earlier version of this essay appeared at the beginning of my 2013 book: Donna Laframboise, *Into the Dustbin: Rajendra Pachauri, the Climate Report & the Nobel Peace Prize* (Port Dover: Ivy Avenue Press, 2013).

2. See, for example, "Climate expert says more extreme weather likely," *Australian Broadcasting Corporation*, 10 February 2011, accessed 11 July 2014, at http://www.abc.net.au/news/2010-12-31/climate-expert-says-more-extreme-weather-likely/1891882, and "Climate expert says more extreme weather likely," *Australian Broadcasting Corporation*, 31 December 2010, at https://www.youtube.com/watch?v=A1U4zXd-FN0.

3. See D. Maraniss and E. Nakashima, "Gore's Grades Belie Image of Studiousness: His School Transcripts Are a Lot Like Bush's," *The*

Washington Post, 19 March 2000, accessed 11 July 2014, http://www.
webcitation.org/69OartUCO.

4. "iDEA Conference April 6-7, Adelaide," *Australian Medical Students'
Association*, 15 March 2013, accessed 11 July 2014, https://www.amsa.
org.au/uncategorized/20130315-idea-conference/.

5. "Carlton conversations @ the Clare Castle Hotel: 2011 series—'What I
believe and why,'" *Church of All Nations*, accessed 11 July 2014, http://
carlton-uca.org/australia-dreaming/conversations.php.

6. "Julie Arblaster," accessed 11 July 2014, http://www.cgd.ucar.edu/ccr/jma/
cv2013.pdf.

7. "Nobel winner says 'important' for IPCC to see landmark climate snapshots," *NSW Department of Primary Industries*, accessed 11 July 2014, http://www.dpi.nsw.gov.au/archive/
agriculture-today-stories/ag-today-archive/june-2012/
nobel-winner-says-important-for-ipcc-to-see-landmark-climate-snapshots.

8. Brisbane City Council, *Brisbane Meeting Planners' Guide, 2011-2013*
(Brisbane: Brisbane Marketing, 2011), 13.

9. Amazon.com, "Return to Almora [Kindle Edition]," accessed July 11,
2014, http://www.amazon.com/Return-Almora-R-K-Pachauri-ebook/dp/
B00DBTHBZY.

10. District of Columbia, "Michael D. Mann v National Review Inc," p2,
accessed 19 Novmeber 2014, http://legaltimes.typepad.com/files/michael-
mann-complaint.pdf.

11. Ibid, p6.

12. Ibid, p3.

13. IPCC Statement: *Statement about the 2007 Nobel Peace Prize*, issued
December 2012, accessed19 Novmeber 2014, http://www.ipcc.ch/pdf/
nobel/Nobel_statement_final.pdf.

14. Letters, "Tusk-Tusk", *The Walrus,* from the June 2013 Magazine, accessed
19 November 2014 http://thewalrus.ca/letters-v10-n5/

15. For the article in question, see "Professor Mann claims to win Nobel

Prize; Nobel Committee says he has not," *examiner.com*, 26 October 2012, accessed 7 July 2014, http://www.examiner.com/article/professor-mann-claims-to-win-nobel-prize-nobel-committee-says-he-has-not.

Chapter 18
The scientists and the apocalypse

1.   R. Darwall, *The Age of Global Warming: A History* (London: Quartet Books, 2013).

2.   United States Committee for the Global Atmospheric Research Program, *Understanding climatic change: a programme for action* (Washington, D.C.: National Academy of Sciences, 1975); Australian Academy of Science Committee on Climatic Change, *Report of a committee on climatic change* (Canberra: Australian Academy of Science, 1976); World Meteorological Organisation, "WMO Statement on Climatic Change," *WMO Bulletin*, Vol. 25 No. 3 (1976), 211-2.

3.   National Research Council, "Ad Hoc Study Group on Carbon Dioxide and Climate," *Carbon Dioxide and Climate: A Scientific Assessment / Report of an Ad Hoc Study Group on Carbon Dioxide* (Washington, D.C.: National Academy of Sciences; 1979); *Proceedings of the Carbon Dioxide and Climate Research Program Conference* (Washington, D.C., 1980); National Research Council, *Changing climate: report of the Carbon Dioxide Assessment Committee* (Washington, D.C: National Academy Press, 1983).

4.   International Council of Scientific Unions Scientific Committee on Problems of the Environment, *SCOPE 29: The Greenhouse effect, climatic change, and ecosystems* (Wiley, 1986).

5.   *Report of the International Conference on the Assessment of the Role of Carbon Dioxide and of Other Greenhouse Gases in Climate Variations and Associated Impacts, Villach, Austria, 9-15 October 1985* (Geneva: World Meteorological Organization, 1986).

6.   International Council of Scientific Unions Scientific Committee on

Problems of the Environment, *SCOPE 29*, as above.

7. W. C. Clark ed., *Carbon dioxide review: 1982* (Oxford: Oxford University Press, 1982); M. C. MacCracken and F. M. Luther, *Detecting the climatic effects of increasing carbon dioxide* (Washington, D.C.: U.S. Dept. of Energy, 1985); R. A. Kerr, "Is the Greenhouse Here?", *Science*, Vol. 239 No. 4840 (1988), 559-561; A. C. Revkin, "Endless summer: living with the greenhouse effect," *Discover*, Vol. 9 (1988), 50-61; R. A. Kerr, "How to Fix the Clouds in Greenhouse Models," *Science*, Vol. 243 No. 4887 (1989), 28-9; R. White, "Greenhouse Policy and Climate Uncertainty," *Bulletin of the American Meteorological Society*, Vol. 70 No. 9 (1989), 1123-7.

8. Intergovernmental Panel on Climate Change, *Climate change: the IPCC scientific assessment* (Cambridge: Cambridge University Press, 1990), 202-3.

9. J. E. Hansen, "The Greenhouse Effect: Impacts on Current Global Temperature and Regional Heat Waves," presented to the US Senate Comittee on Energy and Natural Resources on 23 June 1988, in *The Challenge of Global Warming*, ed. Timothy Wirth, Dean E. Abrahamson (Washington, D.C.: Island Press, 1989), 35-44; P. Shabecoff, "Global Warming Has Begun, Expert Tells Senate," *New York Times*, 24 June 1988, 1; R. A. Kerr, "Hansen Vs the World on the Greenhouse Threat," *Science*, Vol. 244 No. 4908 (1989), 1041-3.

10. Intergovernmental Panel on Climate Change, *Climate change*, as above.

11. T. M. L. Wigley and P. D. Jones, "Detecting CO2-induced climatic change," *Nature*, Vol. 292 No. 5820 (1981), 205.

12. Intergovernmental Panel on Climate Change, *Climate change*, 253, as above.

13. B. J. O'Brien, *Postponing greenhouse: climate change: facts, issues and policies in 1990* (Perth, W.A.: Eco Ethics, 1990).

14. Darwall, *The Age of Global Warming*, as above.

15. Intergovernmental Panel on Climate Change, *Report of the First Session of the WMO/UNEP Intergovernmental Panel on Climate Change (IPCC)* TD 267 (Geneva: World Climate Programme, 1988).

16. United Nations, *General Assembly Resolution on Protection of Global Climate for Present and Future Generations of Mankind* RES/43/53 (1988).

17. B. Bolin, *A History of the Science and Politics of Climate Change: The Role of the Intergovernmental Panel on Climate Change* (Cambridge: Cambridge University Press, 2008), 53-8.

18. *Ibid.,* 54-5.

19. United Nations, *General Assembly Resolution on Protection of Global Climate for Present and Future Generations of Mankind* RES/44/207 (1989).

20. WMO/UNEP Intergovernmental Panel on Climate Change, *Report of the Third Session of the WMO/UNEP Intergovernmental Panel on Climate Change* IPCC 5 (1988).

21. World Meteorological Organization, *WMO Executive Council: Forty-Second Session, Geneva, 11-22 June 1990, Abridged Report with Resolutions* Report No. WMO 739 (1990).

22. M. Thatcher, "Speech at Second World Climate conference," speech presented at Second World Climate conference, Geneva, 1990.

23. Darwall, *The Age of Global Warming*, 52-68, as above.

24. Australia Delegation to the Intergovernmental Panel on Climate Change, *WMO/UNEP Intergovernmental Panel on Climate Change (IPCC), fourth session: Australian Delegation report* (Sundsvall: Sweden, 1990).

25. Intergovernmental Panel on Climate Change, *Climate change: the 1990 and 1992 IPCC assessments: IPCC first assessment report overview and policymaker summaries and 1992 IPCC supplement* (Intergovernmental Panel on Climate Change, 1992), 51-62; See also: Australia Delegation to the Intergovernmental Panel on Climate Change, *WMO/UNEP IPCC, fourth session*; and Bolin, *A History of the Science and Politics of Climate Change*, 67-8.

26.   United Nations, *General Assembly Resolution on Protection of Global Climate for Present and Future Generations of Mankind* RES/45/212 (1990).

27.   United Nations, *Framework Convention on Climate Change* (UN FCCC, 1992).

28.   Intergovernmental Panel on Climate Change Working Group I, *Climate change 1992: the supplementary report to the IPCC scientific assessment* (Cambridge: Cambridge University Press, 1992), 5.

29.   Bolin, *A History of the Science and Politics of Climate Change*, 87-8; see also B. Bolin, "Scientific Assessment of Climate Change," in *International Politics of Climate Change: Key Issues and Critical Actors*, ed. Gunnar Fermann (Oslo: Scandinavian University Press, 1997), 83-109.

30.   F. Pearce, "Frankenstein syndrome hits climate treaty," New Scientist, Vol. 142 Issue 1929 (1994), 5; see also: Bolin, *A History of the Science and Politics of Climate Change*, 86-7.

31.   Ibid.

32.   A. Abbott, "Climate change panel to remain main source of advice," *Nature*, Vol. 374 No. 6523 (1995), 584.

33.   T. Barnett, B. Santer, P. Jones, R. Bradley and K. Briffa, "Estimates of low frequency natural variability in near-surface air temperature," *Holocene*, Vol. 6 No. 3 (1996), 255-63.

34.   B. D. Santer, T. M. L. Wigley, T. P. Barnett and E. Anyamba, "Chapter 8: Detection of climate change and attribution of causes," in *Climate change 1995: the science of climate change,* ed. J. Houghton, L. G. Meiro Filho, B. A. Callander, N. Harris, A. Kattenburg, and K. Maskell (Cambridge: Cambridge University Press, 1996).

35.   W. K. Stevens, "Global Warming Experts Call Human Role Likely," *New York Times*, 10 September 1995, 1; R. A. Kerr, "Scientists see greenhouse, semiofficially," *Science*, Vol. 269, No. 5231 (1995), 1667.

36.   Intergovernmental Panel on Climate Change, *Collated Comments on the Draft Summary for Policy Makers (SPM) of the draft contribution of Working Group I to the IPCC Second Assessment Report submitted by non-government*

organisations before the meeting, WGI/5th/Doc5/26.XI.95 (IPCC, 1995).

37. F. Pearce, "Price of life sends temperatures soaring," *New Scientist,* Vol. 1971 (1995), 5; E. Masood, "Developing-Countries Dispute Use of Figures on Climate-Change Impacts," *Nature,* Vol. 376 No. 6539 (1995), 374-374; A. Meyer, "Economics of Climate-Change," *Nature,* Vol. 378 No. 6556 (1995), 433-433; E. Masood, "Temperature rises in dispute over costing climate change," *Nature,* Vol. 378 No. 6556 (1995), 429.

38. R. A. Kerr, "Studies say—tentatively—that greenhouse warming is here," *Science,* Vol. 268 No. 5217 (1995), 1567-8.

39. P. N. Edwards and S. H. Schneider, "The 1995 IPCC report: Broad consensus or 'scientific cleansing'," *Ecofable/Ecoscience,* Vol. 1 No. 1 (1997), 3-9; J. Zillman, *Third Drafting Session for the IPCC Second Scientific Assessment of Climate Change [Australian Delegation Report]* (Asheville, USA, 1995).

40. J. Houghton, "Meetings that changed the world: Madrid 1995: Diagnosing climate change," *Nature,* Vol. 455 No. 7214 (2008), 737-8.

41. Global Climate Coalition, *The IPCC: Institutionalized 'Scientific Cleansing,' circulated to politicians in Washington by fax machine,* 1996; D. Wamsted, "Doctoring the Documents?" *Energy Daily,* Vol. 24 No. 98 (1995), 1-2.

42. E. Masood, "Climate panel confirms human role in warming, fights off oil states," *Nature,* Vol. 378 No. 6557 (1995), 524; N. Hawkes, "The human role in climatic change," *The Times,* 24 June 1996; S. J. Houghton, *In the Eye of the Storm* (Oxford: Lion Books, 2013).

43. C. Macilwain, "Climate critics claim access blocked to unpublished data," *Nature,* Vol. 378 No. 6555 (1995), 329; T. A. Wigley, "A successful prediction?" *Nature,* Vol. 376 No. 6540 (1995), 463.

44. B. Bert and J. Houghton, "Berlin and Global Warming Policy," Nature, Vol. 374 No. 6519 (1995), 199-200; "Global Warming Rows," *Nature,* Vol. 378 No. 6555 (1995), 322-322.

45. "Climate debate must not overheat," *Nature,* Vol. 381 No. 6583 (1996), 539.

46. A. Montford, *Nullius in verba: the Royal Society and climate change* (London: The Global Warming Policy Foundation, 2012), 38.

47. "Consensus: 97% of climate scientists agree," NASA, accessed 5 May 2014, http://climate.nasa.gov/scientific-consensus

Chapter 19
The scientific method (and other heresies)

1. D. Jones, "Our hot, dry future," *The Age*, 6 October 2008, accessed 10 July 2014, http://www.theage.com.au/federal-politics/our-hot-dry-future-20081005-4udg.html.

2. R. Macey, "This drought may never break," *The Sydney Morning Herald*, 4 January 2008, accessed 10 July 2014, http://www.smh.com.au/news/environment/this-drought-may-never-break/2008/01/03/1198949986473.html.

3. M. Fyfe, "It's not drought, it's climate change, say scientists," *The Age*, 30 August 2009, accessed 10 July 2014, http://www.theage.com.au/national/its-not-drought-its-climate-change-say-scientists-20090829-f3cd.html.

4. D. Karoly, J. Risbey and A. Reynolds, *Climate Change—Global Warming Contributes to Australia's Worst Drought* (Sydney: World Wildlife Fund, 14 January 2003).

5. Ibid, 1.

6. N. Lockart, D. Kavetski and S. W. Franks, "On the role of soil moisture in daytime evolution of temperatures," *Hydrological Processes: An International Journal*, Vol. 27 No. 26 (2013), 3896-3904.

7. W. J. Cai and T. Cowan, "Evidence of impacts from rising temperature on inflows to the Murray-Darling Basin," *Geophysical Research Letters* Vol. 35 (2008).

8. See, for example, A. Dai, K. E. Trenberth, T. Qian, "A global data set of Palmer Drought Severity Index for 1870-2002: relationship with soil moisture and effects of surface warming," *Journal of Hydrometeorology*, Vol. 5 (2004), 1117-1130; K. R. Briffa, G. van der Schrier and P. D. Jones, "Wet and dry summers in Europe since 1750: evidence of increas-

ing drought," *International Journal of Climatology* Vol. 29, No. 13 (2009), 1894-1905.

9. J. Sheffield, E. F. Wood and M. L. Roderick, "Little change in global drought over the past 60 years," *Nature ,*Vol. 491 (2012), 435-438, 435.

10. K. E. Trenberth, "Framing the way to relate climate extremes to climate change," *Climatic Change*, Vol. 115 (2012), 283-290.

11. S-K. Min, X. Zhang, F. W. Zwiers and G. C. Hegerl, "Human contribution to more intense precipitation extremes," *Nature*, Vol. 470 (2011), 378-381.

12. P. W. Stackhouse, S. K. Gupta, S. J. Cox, T. Zhang, J. C. Mikovitz, L. M. Hinkelman, "24.5-Year Surface Radiation Budget Data Set Released," *GEWEX News,* Vol. 21 No. 1 (February 2011), accessed 10 July 2014, http://www.gewex.org/images/Feb2011.pdf.

13. For example, see M. Wild, "Global dimming and brightening: A review," *Journal of Geophysical Research*, Vol. 114, No. D10 (2009).

14. M. H. England, S. McGregor, P. Spence, G. A. Meehl, A. Timmermann, W. Cai, A. S. Gupta, M. J. McPhaden, A. Purich and A. Santoso, "Recent intensification of wind-driven circulation in the Pacific and the ongoing warming hiatus," *Nature Climate Change*, Vol. 4 (2014), 222-227.

15. N. J. Mantua, S. R. Hare, Y. Zhang, J. M. Wallace and R. C. Francis, "A Pacific decadal climate oscillation with impacts on salmon, *Bulletin of the American Meterological Society*, Vol. 78 (1997), 1069-1079.

16. S. Power, T. Casey, C. Folland, A. Colman and V. Mehta, "Interdecadal modulation of the impact of ENSO on Australia," *Climate Dynamics*, Vol. 15 No. 5 (1999), 319-324.

17. See, for example, S. W. Franks, "Assessing Global climate scenarios for hydrological impact assessment," *Hydro 2000: Interactive Hydrology, Proceedings* (Canberra: Institution of Engineers, 2000); S. W. Franks, "Assessing hydrological change: deterministic general circulation models or spurious solar correction? *Hydrological Processes*, Vol. 16, No. 2 (2002), 559-564; A. S. Kiem, S. W. Franks and G. Kuczera, "Multi-decadal vari-

ability of flood risk," *Geophysical Research Letters*, Vol. 30 No. 2 (2003); D. C. Verdon and S. W. Franks, "Long-term behaviour of ENSO: Interactions with the PDO over the past 400 years inferred from paleoclimate records," *Geophysical Research Letters*, Vol. 33 No. 5 (2006), among many others.

Chapter 20
Extreme weather and global warming

1. J. A. Church and N. J. White, "A 20th century acceleration in global sea-level rise," *Geophysical Research Letters*, Vol. 33, No. 1 (2006), accessed 14 July 2014, http://onlinelibrary.wiley.com/doi/10.1029/2005GL024826/pdf.

2. R. J. Nicholls and A. Cazenave, "Sea-Level Rise and Its Impact on Coastal Zones," *Science*, Vol. 328 No. 5985 (2010), 1517-1520.

3. A. Leiserowitz, G. Feinberg, S. Rosenthal, N. Smith, A. Anderson, C. Roser-Renouf and E. Maibach, *What's In A Name? Global Warming vs. Climate Change* (New Haven: Yale University and George Mason University, 2014).

4. J. P. Holdren, *Global Climate Disruption: What Do We Know? What Should We Do?* (Cambridge: John F. Kennedy School of Government, Harvard University, 2007), accessed 10 July 10 2014, http://belfercenter.ksg.harvard.edu/publication/17661/global_climate_disruption.html.

5. J. Hansen, M. Sato and R. Ruedy, "Perception of climate change," *Proceedings of the National Academy of Sciences of the United States of America*, Vol. 109, No. 37 (2012), accessed 10 July 2014, http://www.pnas.org/content/109/37/E2415.full.pdf+html%20target=_blank.

6. P. C. Knappenberger, "Unloading James Hansen's Climate Dice," *Watts Up With That?*, 1 October 2012, accessed 11 July 2014, http://wattsupwiththat.com/2012/10/01/unloading-james-hansens-climate-dice/.

7. D. Bouziotas, G. Deskos, N. Mastantonas, D. Tasknias, G. Vangelidis, S. M. Paplexio and D. Koutsoyiannis, "Long-term properties of annual

maximum daily river discharge worldwide," *European Geosciences Union General Assembly 2011, Geophysical Research Abstracts*, Vol. 13 (2011).

8.   IPCC, "Summary for Policymakers," *Climate Change 2013: The Physical Science Basis. Contribution of Working Group I to the Fifth Assessment Report of the Intergovernmental Panel on Climate Change* (Cambridge: Cambridge University Press, 2013), 5, accessed 10 July 2014, http://www.climatechange2013.org/images/report/WG1AR5_SPM_FINAL.pdf.

9.   C. B. Field, V. Barros, T. F. Stocker, Q. Dahe, D. J. Dokken, K. L. Ebi, M. D. Mastrandrea, K. J. Mach, G-K. Plattner, S. K. Allen, M. Tignor and P. M. Midgley, *Managing the Risks of Extreme Events and Disasters to Advance Climate Change Adaptation: Special Report of the Intergovernmental Panel on Climate Change* (Cambridge: Cambridge University Press, 2012), 268-269.

10.   Editorial, "Extreme weather: better models are needed before exceptional events can be reliably linked to global warming," *Nature*, Vol. 489 No. 7416 (2012), 335-336.

The Australian publication of Climate Change: The Facts *was supported by many friends of the Institute of Public Affairs, including the following:*

LEN AINSWORTH, STEPHEN AINSWORTH, ROB ALBON,
MICHAEL ALDRED, DICK ALLPASS, STEVE ANTONIO,
DAVID ARMSTRONG, ANDREW BADAGOFF, MALCOLM BADGERY,
JANET BARLOW, CHARLES BARNES, BRIAN BEDKOBER, REX BEHAN,
ALBERT BENSIMON, MICHAEL BENSON, RITA & DAVID BENTLEY,
GEORGE BINDLEY, DONALD BLANKSBY, PETER BLOINK,
DES BOLSTER, TOM BOSTOCK, DAVID BOYD, IAN BRACKENRIDGE,
TOM BRINKWORTH, GEOFF BROWN, PETER BRUN, IAN BUNCE,
KEN BURR, ANDY BUTTFIELD, ANNE BUXTON, JOHN BYKERK,
DAVID CAMPBELL, HUME K CAMPBELL, MICHAEL CANE,
CAPITAL CLIMATE ROUND TABLE RATIONALISTS,
LEN & WENDY CARLSON, PETER CARPENTER, PAUL CARR,
JOHN S CHAMBERS, PETER CHAMPNESS, PAUL CHAPMAN,
BOB CHARLES, GEORGE CHOMLEY, JOCK CLOUGH, ALAN H COLE,
OWEN COLTMAN, WALTER & CHRISTL COMMINS,
SIMON COOLICAN, JOHN CORDUKES, JOHN CORRIGAN,
HUGH COWLING, DAVID J CRAZE, MEL CROSS, ALEX CRUICKSHANK,
FRANK CUFONE, LAWRENCE CUMMINGS, DOUGLAS CUSTANCE,
BRUCE CUTTS, KEVIN DALY, BRYCE DAVEY, IRENE DAVIES,
BRENDON DAVIS, TONY DEAR, HARALD DEN HARTOG,
KEITH DENNIS, MICHAEL DINGSDALE, JOHN DOHERTY,
JAMES DOOGUE, RUSSELL DOWLING, JERRY DRAGOSEVIC,
AERT DRIESSEN, TIM DUNCAN, MIKE ELLIOTT, ANDREW EVANS,
PETER FARRELL, NEIL FEARIS, TIMOTHY FITZPATRICK,
IAN FLANIGAN, JIM FLETCHER, NICHOLAS FORD, PETER FRANKLIN,
ROSS A FRICK, JANET FRICKE, MICHAEL FRY, JAN GIBBONS,
ROGER GIBBONS, DAVID GIBBS, BRIAN GIBSON, BRETT GILL,
CHRISTOPHER GOLIS, ADRIAN GOOD, DAVID GRACE, BOB GRAHAM,
WARRINE PASTORAL CO., MICHAEL GRAY, RICHARD GRAY,
MARK GRIGOLEIT, EDWIN GRIMSHAW, IAN GUTHRIE,
CHRISTOPHER HALL, ROD HALLIDAY, TOM HARLEY,
CHRIS HARRINGTON, GRAEME HAUSSMANN, SHAUN HAZELL,
PETER VANRENEN, MARK HEALD, IAN HILL, TONY HODGSON,
MICHAEL HOGG, GEOFF HONE, SHAMUS HOWDEN, OWEN HUGHES,
STEPHEN D HUGHES, DAVID H HUME, BRIAN J HURLOCK,
HARRY IMBER, ROHAN INGLETON, GRAEME INKSTER, ALAN IRVING,

MURRAY JAMIESON, LYN JEFFRIES, KEN JEPPESEN, LES JOHNSON,
ALAN JONES, BRIAN JONES, STEPHEN JONES, TONY KEENAN,
ROD KEMP, MALCOLM KENNEDY, JOHN KOPCHEFF,
MICHAEL KROGER, GLEN LAMPERD, STEFAN LANDHERR,
PHILLIP LAW, ALFRED LEDNER, PAUL LEMM, SAMUEL J,
GEORGE LLOYD, KEVIN R LOHSE, KEVIN LORENZ, MATTHEW LUCAS,
RICHARD MALLETT, CLEMENT MARRINAN, MIRRIE MASNJAK,
ROBERT MASTERMAN, JOCELYN MAXWELL, W R MAXWELL,
JEFF MCCLOY, TERRY MCCRANN, GARY MCGILL, GREG MCKAY,
MALCOLM MCKELLAR, TONY MCLEAN, AARON MEAD,
W G H MEADOWS, PAUL MICHAEL, CRAIG MILLS, ROD MILTON,
KATHLEEN MILTON, MICHAEL & SUE MINSHALL,
COLMAN MOLONEY, JOHN MONTGOMERY, ROB MOORE,
HUGH MORGAN, RICHARD MORGAN, ROBERT MORGAN,
DAVID MORRIS, KARL MORRIS, JOHN NETHERY, GRAEME NICOL,
PETER NIXON, TERRY O'CONNOR, COLIN O'NEIL, JANETTE O'NEIL,
RODNEY & JUDITH O'NEIL, BERNARD O'SHEA, ANTHONY PARKER,
GEOFF PARKER, LES PARSONS, MAX PARSONS, PETER PATISTEAS,
JOHN PENNINGTON, RICHARD PEPPARD, JONATHON PERKS,
PETER PHELPS, SIMON PHIPPS, JOHN QUINLAN, LACHLAN REEKS,
ANTHONY REID, REYMOND REINECKE, EVAN RICHARDS,
ERMANNO RIHS, JASON RONALD, SIMON RUSZCZAK, TONY SAGE,
MURRAY SANDLAND, STEPHEN SASSE, MICHAEL SCADDING,
GREG SCHOFIELD, GRAHAM SELLARS-JONES, MICHAEL SHIRLEY,
MARIO D SIMIC, DOUGLAS SPENCE, EVE ELIZABETH SPENCE,
HELEN STAHMANN, MICHAEL STEYN, HAMISH STITT,
STOCKHOLMS INITIATIVET, NEIL STUART, ROD STUART,
YVONNE SUMNER, VALERIE TAYLOR, BILL THOMAS,
PAUL THOMAS, JAMES THOMSON, MICHAEL TRIFUNOVIC,
NOEL TURNBULL, ALIX TURNER, EWEN TYLER, DAVID VAN GEND,
ROD WALKE, JOHN WALKER, BARRY WALSH, DAVE WANE,
JENNESS WARIN, ROBERT WATT, GORDON WHITE, JOHN WHITE,
KYLE WIGHTMAN, GARTH WILLIAMS, T ANDREW K WILSON,
PHILIP R WOOD, and JOHN WYLD